# ECOLOGICAL FORECASTING

# ECOLOGICAL FORECASTING

**Michael C. Dietze**

PRINCETON UNIVERSITY PRESS

PRINCETON AND OXFORD

Published by Princeton University Press, 41 William Street,
Princeton, New Jersey 08540

In the United Kingdom: Princeton University Press, 6 Oxford Street,
Woodstock, Oxfordshire OX20 1TR

press.princeton.edu

Jacket art courtesy of the author and Shutterstock

Library of Congress Cataloging-in-Publication Data

Names: Dietze, Michael Christopher, 1976– author.
Title: Ecological forecasting / Michael C. Dietze.
Description: Princeton : Princeton University Press, [2017] | Includes
  bibliographical references and index.
Identifiers: LCCN 2016044327 | ISBN 9780691160573 (hardcover : acid-free
  paper)
Subjects: LCSH: Ecosystem health—Forecasting. | Ecology—Forecasting.
Classification: LCC QH541.15.E265 D54 2017 | DDC 577.01/12—dc23
  LC record available at https://lccn.loc.gov/2016044327

British Library Cataloging-in-Publication Data is available

This book has been composed in Sabon

Printed on acid-free paper. ∞

Printed in the United States of America

10 9 8 7 6 5 4 3 2 1

# Contents

## Preface

THIS BOOK, AS well as my broader interest in ecological forecasting, is the result of a bit of academic serendipity. In the summer of 2001, less than a year into grad school and at that perfect stage for latching onto an idea, Clark et al. (2001) published "Ecological Forecasts: An Emerging Imperative," arguing that to make ecology more relevant to society, and the problems we claim to care about, requires that we make our discipline more predictive. The young, nature-loving, fledgling ecologist in me and the quantitatively focused former engineering student in me both embraced this call to action. By a bit of luck the lead author happened to be my graduate advisor, and the rest, so they say, is history.

Or is it? In the intervening years since Clark et al. (2001), to what degree has ecological forecasting emerged as a distinct discipline unto itself? Is the supply of ecological forecasts keeping up with the demand? Is ecology becoming a more predictive science? I believe the time is ripe to assess where we have come as a field. What have been the successes of ecological forecasting, and what is holding us back? Having spent the last 15 years working and teaching in this area, I am convinced one of the things holding us back is a shortage of training in the set of statistical, modeling, and informatics tools necessary to tackle the challenges before us in an era of big data and even bigger environmental problems. This book aims to help fill that gap.

In many ways this book describes an emerging new way of doing science in ecology that involves a much closer connection between models and data. As such the skills and techniques in an ecological forecaster's toolbox include, but go beyond, those of a traditional modeler's. Forecasting a real place and time is a much more data-intensive process than theoretical modeling, and draws heavily upon skills from statistics and informatics. At the same time it relies on more sophisticated modeling than is often employed for pure data analysis. Indeed, most of this book is focused on the statistics and informatics of model-data fusion and data assimilation. The literature on these topics has largely been scattered across other disciplines, and adoption by ecologists has been encumbered by discipline-specific jargon and mathematical notation that is difficult to penetrate. This has been unfortunate, because while many of these techniques require a change in how one perceives a problem, they are not unreasonably complicated or difficult to implement.

As noted earlier, the backbone of this book covers topics related to the statistics and informatics of model-data fusion: data management, workflows, Bayesian statistics, uncertainty analysis, fusing multiple data sources, assessing model performance, and a suite of data assimilation techniques. Throughout these chapters the aim is to provide enough technical detail to enable the reader to understand how the methods work and to be able to implement them. Subtleties will be highlighted when they will help readers avoid common pitfalls or misinterpretations, but otherwise the goal is for the text to remain accessible and not get bogged down in esoteric details. The focus is very much on application not theory. That said, *this is definitely a book about concepts, not a training manual.*

Interspersed among these concept chapters are "case study" chapters on specific ecological subdisciplines that aim to highlight the successes and failures of ecological forecasting in each: natural resources, endangered species, epidemiology, and the carbon cycle. These chapters also aim to make it clear that forecasting is useful across broad swaths of ecology, and not limited to specific subdisciplines. Indeed, to date expertise in many of the topics outlined here has been distributed across subdisciplines, with some fields further along in adopting one set of approaches, while another field might excel in some other area. The book also begins and ends with a discussion of decision support, as improving decision making is often the primary driver of many ecological forecasts.

Finally, as with all technical topics, we inevitably need to "get our hands dirty" to really understand what's going on. To facilitate this hands-on learning, while also aiming to keep material up to date, I have made code and tutorials available online as supplemental material. Rather than providing static code within the book itself, the intention is to keep this material up to date and publicly available as the field evolves, something that would not be possible if this book was tied to specific code. This material draws heavily from my graduate course on this same topic and my experience teaching this material at a variety of workshops and summer courses. All code and tutorials are currently hosted at https://github.com/EcoForecast. In addition, this site includes a second repository dedicated to the code used to generate the figures, tables, and analyses described in this book. More generally, the site ecoforecast .org provides information not only about this book and its tutorials but also other literature, workshops, and new developments in the field.

## Acknowledgments

THIS BOOK WOULD not have happened without the tireless support, feedback, and encouragement of Jason McLachlan. The text herein benefited greatly from comments provided by colleagues, my lab, and the students in the Ecological Forecasting courses at Boston University and Notre Dame: Colin Averill, Bethany Blakely, Elizabeth Cowdery, Martha Dee, Meredith Doellman, Hollie Emery, Istem Fer, Diana Gergel, Dan Gianotti, Brittany Hanrahan, Lizzy Hare, Ryan Kelly, Ji Hyun Kim, Kelsey Kremers, Zhan Li, Josh Mantooth, Jackie Hatala Matthes, Afshin Pourmokhtarian, Ann Raiho, A. J. Reisinger, Angela Rigden, Christine Rollinson, Alexey Shiklomanov, Arial Shogren, Toni Viskari, and John Zobitz. I would also like to thank Dave Schimel and an anonymous reviewer for their thoughtful and constructive reviews. My thoughts on the topics covered in this book have developed over they years thanks to constructive interactions with numerous people across the research community. While I couldn't possibly list everyone by name I want to particularly acknowledge the faculty and students on the Flux Course (http://fluxcourse.org/), the FORECAST Research Coordination Network (NSF RCN, #0840964), and the Clark Lab at Duke, each of which would not have happened without the leadership of Dave Moore, Yiqi Luo, and Jim Clark, respectively. Thanks to my editor, Alison Kalett, for seeing potential in this project and giving me the flexibility to work around a tenure clock, and to all of the staff at Princeton University Press who shepherded this through from start to finish. My work on this book was directly supported by NSF Advances in Bioinformatics grants 1062547 and 1458021 and leveraged research under NSF 1241891, 1261582, 1318164, and NASA 13-TE13-0060. Finally, thanks to the Massachusetts Bay Transit Authority's commuter rail system, whose poor Wi-Fi service provided me hundreds of hours of uninterrupted writing time.

# ECOLOGICAL FORECASTING

# 1

## Introduction

### 1.1 WHY FORECAST?

Humanity has long depended on natural resources and ecosystem services to survive. However, those same natural systems are increasingly becoming dependent upon humanity for their survival. We live in an era of rapid and interacting changes in the natural world: climate is changing; atmospheric $CO_2$, $CH_4$, $N_2O$, and $O_3$ are rising; land-use change is making habitats smaller and more fragmented; native species are shifting their ranges; exotic species are invading; new diseases are emerging; and agriculture, forestry, and fisheries are depleting resources unsustainably. Driving all these changes is human population, which has grown by a billion people every 12 to 13 years since the 1970s, and is projected to keep rising (United Nations 2014). That's equivalent to adding another US population to the world every 4 years. Furthermore, the economic activities of humankind have changed radically, with more and more of the world enjoying a higher, but often more resource-intensive, standard of living.

Within this context, decisions are being made every day, at levels from individuals to nations, that affect the maintenance of biodiversity and the sustainability of ecosystem services. Furthermore, the ecological questions being asked by policymakers, managers, and everyday citizens *are fundamentally about the future*. They want to know what's going to happen in response to decisions versus what's going to happen if they do nothing. Those decisions are being made with or without the input of ecologists, and far too often these questions are not being answered with the best available science and data. Ecologists are being asked to respond to unprecedented environmental challenges. So how do we, as ecologists, provide the best available scientific predictions of what will happen in the future? In a nutshell, how do we make ecological forecasts?

The foundational paper "Ecological Forecasting: An Emerging Imperative" (Clark et al. 2001) defines ecological forecasting as

> the process of predicting the state of ecosystems, ecosystem services, and natural capital, with fully specified uncertainties, and is contingent on explicit scenarios for climate, land use, human population, technologies, and economic activity.

However, the knowledge and tools for making ecological forecasts lies outside the training of most ecologists. This information is scattered across the literature, with different ecological subdisciplines often unaware of progress being made in others, and with many forecasting techniques being coopted from other disciplines, such as meteorology and statistics. This book aims to synthesize this literature, both distilling key concepts and highlighting case studies from different ecological disciplines.

The importance of learning to make ecological forecasts goes beyond decision making. The ability to make quantitative, testable predictions, and then confront them with data, is at the heart of the scientific method and advances our basic scientific understanding about ecological processes. In general, a successful forecast into the future, to a new location, or under novel conditions provides much stronger support for a hypothesis than its success in explaining the original data used to develop it. Indeed, it is possible for alternative models to produce similar fits to data, but make very different predictions. By making frequent forecasts, checking those forecasts against data, and then updating our projections, we have the potential to accelerate the pace of our science. Indeed, a key attribute of science is the emphasis on updating our ideas as new information becomes available. Forecasting puts this idea front and center, updating projections routinely as new information becomes available. Furthermore, reframing an ecological question into a forecasting problem can often allow the problem to be seen in a new light or highlight deficiencies in existing theories and models.

Despite the importance of forecasting to making our science more relevant and robust, there are ecologists who will rightfully point out that in ecology our understanding of many processes is often coarse, data are noisy, and dynamics can be idiosyncratic, varying from system to system or site to site in ways that defy our current understanding. These are all valid points, and because of that, much of this book will focus on quantifying, partitioning, and propagating uncertainties and sources of variability. These challenges also highlight the need to update projections in light of new data—for example, adjusting projections after low-probability events, such as disturbances, occur. Indeed, making ecology a more predictive science will rely heavily on understanding probability and uncertainty—ecological forecasts need to be probabilistic to capture these uncertainties.

As ecologists we appreciate the complexities and idiosyncrasies of the systems we study, and at times the multitude of possible interactions and outcomes can seem overwhelming. When faced with questions about how ecological systems will respond to change, it is far too easy to answer these questions with "it depends." However, while the idea that everything in an ecosystem is connected to everything else makes for a profound and deep mythos, the reality is that not all things are connected equally. The responses of complex systems are indeed capable of surprises, but more often their responses are driven by a small subset of interactions and processes. Fundamentally, I believe that ecology is more than just a collection of case studies and just-so stories. The extent to which ecological systems will prove to be forecastable remains a critical, but ultimately empirical, question. One that at the moment is unanswered, but profoundly important to the future of our science and its relevance to society. How we might go about answering that question is the subject of this book.

Figure 1.1 illustrates a general conceptual workflow for how we might develop ecological forecasts and maps the chapters of this book, and how they relate to one another, onto that workflow. In brief, making forecasts ultimately depends on models (center), but models are dependent on data in many ways (chapters 3, 4). Data are used as real-world drivers, constrain parameters (chapters 5, 6, 8, 9), and provide observations of system states to both initialize (chapters 13, 14) and validate (chap-

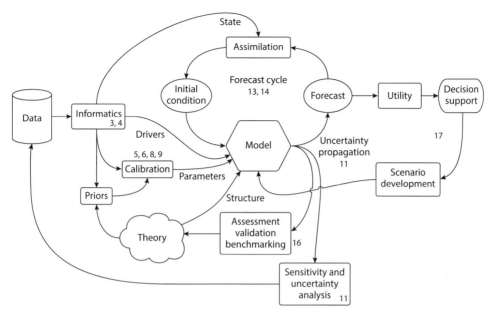

FIGURE 1.1. Ecological Forecasting conceptual workflow. Relevant chapters identified by number.

ter 16) models. Uncertainties enter at all stages, and need to be partitioned into different sources (chapter 6) and propagated into forecasts (chapter 11). Improving forecasts relies critically upon a number of feedback loops: sensitivity and uncertainty analyses (chapter 11) identify key parameters and processes driving uncertainties; model assessment, validation, and benchmarking (chapter 16) identify model structural errors; and data assimilation, through the forecast cycle (chapters 13, 14), addresses model process error. While some forecasts are done to advance basic science, many are done to provide explicit decision support (chapter 17), evaluating specific scenarios and allowing decision makers to assess trade-offs among alternatives.

Beyond this conceptual workflow, there are a number of major themes cutting across the book, which are summarized in the following sections.

## 1.2 THE INFORMATICS CHALLENGE IN FORECASTING

Ecological forecasts are generally focused on making projections for specific places and systems in the real world. This focus on the real world means that much of the day-to-day effort in producing ecological forecasts may end up being devoted to data management and processing. Since we generally want to update forecasts as new data become available, this puts a premium on the documentation and automation of workflows. Furthermore, since many forecasts are developed explicitly for policy and management goals, forecasts must be transparent, repeatable, and adhere to best practices. Ultimately, up-front attention to the informatics challenges of forecasting not only makes for better science, but, by reducing error and increasing automation, it also can allow you to do more science. Unfortunately, these informatics challenges are outside the core training of many ecologists. While the field of informatics is rapidly evolving, with new tools being developed daily, many of the core concepts are persistent. Thus the informatics of forecasting is discussed up front in this book (chapters 3 and 4) as a foundation that will be leveraged when performing syntheses and making forecasts.

## 1.3 THE MODEL-DATA LOOP

In response to the multitude of changes occurring in the natural world, ecologists have invested considerable effort in monitoring. Ecological monitoring occurs across a wide range of scales and systems, from individual organisms being tagged to satellites measuring the globe, and has become a fundamental mission of many federal agencies and nongovernmental organizations (NGOs) around the world. However, *the core aim of monitoring is to detect change after it has happened, rather than to anticipate such change*. As such, monitoring is a primarily data-driven exercise, and can actually occur in a relatively model-free manner. By contrast, forecasting, by its nature, requires that we embrace models, as they are our only way to project our current understanding into the future, to new locations, or into new conditions.

Because they are the backbone of forecasting, I will be discussing models throughout this book. If the thought of that makes you anxious, I'll ask you to stick with me because this is most definitely not a book about modeling. Furthermore, models don't have to be complex and impenetrable. Even basic statistical models, such as ANOVAs and regressions, are models, the former predicting that different groups have different means and the latter predicting a straight line. For almost everything I discuss in this book, *the underlying principles are the same regardless of how simple or complex the model*. Indeed, I will often rely on very simple models, such as fitting a mean, to illustrate concepts that are applicable to all models regardless of complexity.

Across this spectrum of model complexity, the common denominator is that models remain a quantitative distillation and formalization of our hypotheses about how a system works. Because they embody our current working hypotheses, models are relevant to all ecologists, not just modelers and theoreticians. Unfortunately, in ecology there has often been a disconnect between the empirical and modeling portions of the community. In truth, all ecologists are quantitative ecologists, because we all use numbers to tell things apart. Similarly, any ecologist who has ever done statistics is a modeler. Or better yet, we should recognize than none of us are "modelers" any more than a geneticist is a "PCRer"—models are tools we all use, not a separate discipline, and we're all just ecologists.

While models are a critical part of any forecast, so are data. Unlike many theoretical modeling exercises, forecasts focus on projecting real systems, and working in real systems requires data about those systems. Therefore, *any approach to forecasting, from the most simple to the most complex, requires the combination of models and data*. In combining models and data, forecasting places a premium on quantifying the uncertainties in both. The estimation of uncertainty in a forecast is critical to decision support, as underestimating uncertainty can lead to overconfident decisions, while overestimating uncertainty can lead to excessive, and expensive, levels of caution. Once uncertainties are quantified, we will sometimes find that effective decision making (or hypothesis testing) requires that these uncertainties be reduced. The next job of the forecaster is thus to identify the sources of new data that will have the greatest or most cost-effective impact on reducing model uncertainty.

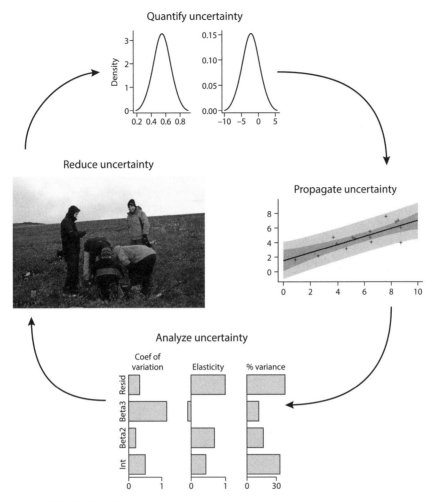

FIGURE 1.2. Model-data loop.

Throughout this book I will talk about the idea of the model-data loop (figure 1.2). This refers to the idea of iteratively using data to constrain models and then using models to determine what new data is most needed. The model-data loop applies regardless of the complexity of the model, and indeed should play a central role in any decision to increase model complexity. As an example, consider a simple linear model, $Y = \beta_1 + \beta_2 X_2 + \beta_3 X_3$, with one dependent variable, $Y$, and two independent variables, $X_2$ and $X_3$. The model-data loop might begin by *quantifying the uncertainty* in the intercept ($\beta_1$), the two slopes ($\beta_2$ and $\beta_3$), and the residual error using regression. Alternatively, it might require more sophisticated techniques to characterize the uncertainty (chapter 6) if there are additional complexities such as non-Normal error, non-constant variance, missing data, or multiple sources of error (for example, observation versus process, errors in the $X$'s). The next step would be to *propagate the uncertainty* in the model into the forecast (chapter 11). For a regression this may be something as simple as constructing confidence and predictive intervals. Next we would *analyze the uncertainty* to discern which processes are controlling our predictive uncertainty.

Figure 1.2 illustrates a case where the intercept and residuals dominate model uncertainty, contributing 42% and 33% respectively to the predictive variance, while the first and second slope terms contribute 15% and 10%. Finally, based on the uncertainty analysis, we could target additional field measurements. In this case we might choose to simply increase the overall sample size, which will better constrain all the regression parameters. Furthermore, since $X_2$ explains more of the variance than $X_3$, sampling may be stratified to better capture variability in $X_2$.

The idea of using models to inform field research has been discussed in many contexts (Walker et al. 2014) but rarely occurs in practice, and when it does it is often in a qualitative way, such as identifying processes that we don't understand (Medlyn et al. 2015). Approaches to uncertainty analysis discussed here aim to make this a more quantitative endeavor (LeBauer et al. 2013; Dietze et al. 2014). While model-driven data collection is no substitute for hypothesis testing, the two very often go hand in hand. We almost always have open research questions about the processes identified as driving forecast uncertainty. More pragmatically, we are likely to have unanswered hypotheses in many areas of our research, but the model-data loop allows us to focus first on those that will give us the most return on investment when allocating scarce resources.

## 1.4 WHY BAYES?

Throughout this book many of the concepts and tools presented will be developed from a Bayesian statistical perspective. At this stage I want to quickly discuss the rationale for this choice and defer a more detailed description of Bayes and its forecasting applications to later chapters (chapters 5, 6, 8, 9, 13, and 14). This book assumes no prior experience with Bayesian approaches—in particular, chapter 5 aims to provide a solid common ground for readers who may not have had a prior exposure to Bayesian concepts and methods. For those with previous experience with Bayes, chapter 5 will serve as a refresher or could be skipped. That said, this is a book on forecasting, not Bayesian statistics, and those with no prior exposure to Bayes will probably find chapter 5 to be accelerated and may want to consult an introductory textbook for additional background (Clark 2007; Hobbs and Hooten 2015).

While it is definitely possible to do ecological forecasting from a classical (frequentist) statistical perspective, the Bayesian approach has a number of advantages that are even more valuable in forecasting than they are in standard data analysis (Ellison 2004; Clark 2005). First, it allows us to *treat all the terms in a forecast as probability distributions*. Treating quantities we are interested in as probabilities makes it easier to quantify uncertainties, to partition these uncertainties into different sources of error and process variability (chapter 6), and to propagate them into forecasts (chapter 11) and decision support (chapter 17). Second, as discussed earlier, the *ability to update predictions as new data becomes available* is a critical aspect of forecasting. In forecasting, we leverage the inherently iterative nature of Bayes' theorem (equation 1.1) as a means of updating forecasts. Specifically, Bayes' theorem states that the updated forecast (also known as the posterior probability) for any quantity, $\theta$, is proportional to the likelihood of the most recently observed data, $y$, times our previous forecast (also known as the prior probability):

$$\underbrace{P(\theta|y)}_{posterior} \propto \underbrace{P(y|\theta)}_{likelihood} \underbrace{P(\theta)}_{prior} \tag{1.1}$$

As new data becomes available, the posterior from the previous forecast becomes the prior for the next. Thus, in forecasting, the priors embody the information provided by previous observations and forecasts. Finally, from a pragmatic perspective, Bayesian numerical methods tend to be flexible and robust, allowing us to deal with the *complexity of real-world data* and to build, fit, and forecast with relatively complex models.

## 1.5 MODELS AS SCAFFOLDS

Traditional approaches to modeling have focused primarily on forward modeling—taking a set of observed inputs and running them through a model to generate a set of outputs (figure 1.3A). The other approach to modeling commonly employed is inverse modeling—starting from a set of observations that correspond to model outputs and trying to infer the most likely inputs (figure 1.3B). In this book I will rely on both of these approaches, but I will also present a third modeling paradigm, which I call "models as scaffolds" (figure 1.3C), which refers to using models and data together to constrain estimates of different ecosystem variables (Dietze et al. 2013). One of the challenges of combining different data sources (chapter 9) is that frequently observations may come from very different spatial and temporal scales, or they may capture different but related processes, and thus cannot be directly compared to one another. In most cases ad hoc approaches, such as interpolating data to a common scale, are not the right answer to this problem, as they throw out uncertainties, misrepresent the actual number of observations, and introduce additional

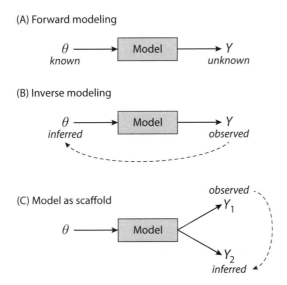

FIGURE 1.3. Modeling approaches. (A) In forward modeling the model inputs and parameters are treated as known and the model is used to predict some unknown state. (B) In inverse modeling the current state is observed (with uncertainty) and the inputs or parameters are inferred. (C) In the model as scaffold approach multiple outputs may be observed and the best estimate of the current state is inferred, both for observed ($Y_1$) and unobserved ($Y_2$) outputs, based on the covariance among model states.

assumptions that may not be consistent with our understanding of the process (for example, linearity, homogeneity, and scale independence). By contrast, it is often the case that the models we use to forecast may represent multiple spatial scales, temporal scales, and interconnected processes. Thus, while different data sets may not be directly comparable to one another, they may all be comparable to the model. In this context combining data and models is a fundamentally data-driven exercise for allowing data sets to talk to each other across different scales and processes. The model is the scaffold on which we hang different observations. Furthermore, this scaffold will leverage the understanding and current working hypotheses that are embedded in the models to describe how we believe different processes and scales are related to one another. In this sense the model serves as a covariance matrix, providing structure to the correlations between different observations. This concept of models as scaffolds will be central not only to how we approach data synthesis but also to how we update forecasts as new observations become available (chapters 13 and 14).

## 1.6 CASE STUDIES AND DECISION SUPPORT

Throughout this book chapters illustrating case studies from different disciplines are interspersed among chapters illustrating tools and concepts. Specifically, the case study chapters look at the status of ecological forecasting in biodiversity and endangered species (chapter 7), natural resource management (chapter 10), the carbon cycle (chapter 12), and disease ecology (chapter 15). No one discipline has solved the ecological forecasting problem, with each being stronger in some areas and weaker in others. While there are differences in the forecasting problem among ecological disciplines, there are also many overarching similarities and a lot of shared experience that is too often isolated in academic silos. These chapters aim to build bridges between these silos.

Finally, in chapter 17, I discuss how ecological forecasts can better inform decision making, touching on approaches to scenario development and quantitative decision support models. While figure 1.1 depicts decision support as an end-goal of ecological forecasting, as indeed it is the motivating factor in many real-world forecasts, in reality there are many feedback loops between forecasting and decision support, with decisions not just providing the scenarios and alternative choices to be evaluated but also defining the context, scope, and desired output variables.

Covering the details of decision support requires that we first develop a strong foundational understanding of how ecological forecasts are generated, which is why decision support is at the end of this book. However, to be able to better frame and contextualize ecological forecasts, it is also important to introduce many of the key concepts of decision support up front. First and foremost of these is that decisions are about what will happen in the future, in response to our choices, rather than about what's already happened in the past—decisions depend upon forecasts. However, decisions also depend upon the presence of choices, which are frequently evaluated as alternatives or scenarios that describe different decisions or storylines about how the external world unfolds. Because of this forecasts will be broadly separated into those that provide *predictions*, probabilistic forecasts based on current trends and conditions, versus those that provide *projections*, probabilistic forecasts driven by explicit scenarios.

Another key concept in decision support is to remember that while science provides facts and knowledge, decisions are ultimately about *values*. Furthermore, it is the values of the community as a whole that are relevant, not just those of the scientist. While it is important to acknowledge that we all have our own values and biases, and that these inevitably affect our research, a key part of formal decision support relies on trying to compartmentalize knowledge and values. We want to objectively determine what the trade-offs are in any particular decision (for example, between different stakeholders or competing goals), so that the relevant decision maker, be that an individual, a committee, or an electorate, can decide how to balance the competing, subjective values in any trade-off. Doing so requires that ecological forecasts have clearly defined *objectives* that summarize what matters to stakeholders. Objectives should also indicate the desired direction of change (for example, increase endangered species population) for individual objectives. These objectives then need to be translated into *performance measures* that quantify the objectives in units appropriate for that objective (number of individuals, habitat area, and so on). It is the future of these performance measures (sometimes dubbed *consequences*), subject to uncertainties and alternatives, that we aim to forecast. Decision support also relies on a broad set of useful *alternatives*—any decision is only as good as the set of choices considered—while also acknowledging that a set of alternatives that is too large will overload both the forecaster and the decision maker.

Fundamentally, the goal of decision support isn't to make "optimal" decisions *on behalf of* the decision maker, but to determine the *trade-offs* among competing performance measures. The only alternatives that should be eliminated by the analyst are those that are strictly dominated (that is, lose-lose on all fronts). Beyond that, determining how stakeholders value the competing objectives in any trade-off may be outside the decision support scope, or may occur through discussion, weighting of different values, or more formal quantification of values using Utility functions (quantitative relationships between a performance measure and value).

When presenting the results of any probabilistic forecast it is important to be aware that humans do not have an innate sense of probability and rely instead on a set of *heuristics* that can be subject to a laundry list of well-known cognitive biases (Kahneman 2013). Because of this, perceptions can be sensitive to how uncertainties are presented. That said, another key concept of decision support is understanding how the *uncertainties in forecasts interact with the inherent risk tolerance or risk aversion of stakeholders*. As mentioned earlier, overestimation of uncertainty can lead to costly levels of caution, while underestimation of uncertainty can lead to risky, overconfident decision making. Even if stakeholders could perceive uncertainties perfectly, these uncertainties would have large impacts on decision making because utility functions are generally nonlinear—we don't perceive losses the same as gains. In general, utility declines as uncertainty increases, which is often dealt with by looking for strategies that are precautionary, robust to uncertainties, or adaptive. Indeed, ecological forecasting is at the core of many adaptive management approaches (Walters 1986).

To conclude, in this chapter I introduced the basic concepts of ecological forecasting and their relevance to both decision making and basic science. Forecasting was shown to depend closely on the integration of models and data, and thus requires a solid foundation in informatics and statistics. In building this foundation, I argued for taking a Bayesian approach to forecasting, as it allows forecasts to be easily updated in the light of new data. I introduced the idea of the "model-data loop," where model analyses are used to identify and prioritize measurements, and the idea of "models as scaffolds" for allowing data of different types and on different scales to be mutually informative. I also introduced some of basic concepts of decision support and how forecasting fits into that framework. Finally, I argued the importance of accounting for uncertainty and that ecological forecasts should be made probabilistically. In the following chapter I will follow up on uncertainty and probability by walking through an example of how different sources of uncertainty affect a simple forecast, using this as a springboard to look more deeply at the problem of predictability in ecology.

## 1.7 KEY CONCEPTS

1. Ecological forecasting is critically important to improving our science, making it more relevant to society, and quantitatively supporting decision making.
2. Forecasting is data-intensive, and there are a number of informatics challenges to making forecasts efficient, repeatable, transparent, and defensible.
3. Forecasting requires combining data with models. Doing so requires the quantification of the uncertainties in both and the propagation of these uncertainties into forecasts.
4. Forecasting concepts apply to all models, from simple statistical models to complex computer simulations.
5. Data inform models, and models can be used to more precisely target future data as part of the model-data loop.
6. Bayesian approaches allow us to represent uncertainties probabilistically and to update forecasts easily as new data becomes available.
7. Models can serve as a scaffold for combining information on different processes, or across different scales in space and time.
8. Ecological forecasting is a key component of decision support, as all decisions are fundamentally about the future and are sensitive to forecast uncertainties. However, decisions are ultimately about human values in how we balance trade-offs among competing objectives.

## 1.8 HANDS-ON ACTIVITIES

https://github.com/EcoForecast/EF_Activities/blob/master/Exercise_01_RPrimer.Rmd

- Primer on the R language

# 2

## From Models to Forecasts

*SYNOPSIS: This chapter starts by looking at how ecological models have traditionally been taught and walks through a simple example of making a prediction using the logistic growth model. The aim of this example is to illustrate the process and challenges of forecasting, the sources of uncertainty in a forecast, and the importance of treating these uncertainties as probability distributions. Building on this example, I discuss what makes some processes predictable, while other forecasts fail, and end by assessing the nature of the ecological forecasting problem from a first-principles perspective.*

### 2.1 THE TRADITIONAL MODELER'S TOOLBOX

In the previous chapter I presented the idea that models play a central role in forecasting because they allow us to project what we know now into the future. However, I'm going to argue that the way most ecologists are taught to think about models is not sufficient to make useful forecasts. Specifically, the core difference between forecasting and theory isn't a specific set of quantitative skills, but rather in the way we see the world, which has been strongly influenced by the deterministic, Newtonian way most of us are taught both science and math.

Like all theory and modeling in ecology, let's start with exponential population growth as an example. The exponential growth model is derived from the simple assumption that per-capita growth rate, $r$, is constant

$$\frac{1}{N}\frac{dN}{dt} = r$$

where $N$ is the population size, $dN/dt$ is the population growth rate over time, and dividing this by $N$ causes this growth rate to be expressed on a per-capita basis. Solving this equation gives the familiar exponential growth equation

$$N(t) = N_0 e^{rt}$$

where $N_0$ is the initial population size at time $t = 0$. This equation has two possible outcomes. If $r$ is negative, then the population declines to extinction. If $r$ is positive, then the population grows exponentially without bound. In terms of traditional approaches to model analysis, we can formalize this and say the exponential growth model has one equilibrium, $N = 0$, and that this equilibrium is stable if $r < 0$ and unstable if $r > 0$.

The preceding discussion highlights how most ecologists are taught about models, and indeed how I have taught many students. This approach emphasizes *equilibria*, the lack of change in a system over time, and *stability*, the tendency for a system to return to equilibrium when perturbed. In more advanced modeling courses, one can spend weeks learning about the equilibria of increasingly complex models, how to solve for them, and how to determine if they are stable. This is usually about the time that my students begin to question me about the value of models. They often see a disconnect between the foci of theory and the real-world systems they want to understand. And here they are correct—traditional ecological theory does a bad job of making *connections between models and measurements*. *Modern ecosystems are rarely in equilibrium*, and *knowing the current trajectory of a system is often more important to policy and management than an asymptotic equilibrium*.

This is not to say that traditional modeling has no value for ecological forecasting. Mathematical analyses of equilibria and stability often give insights into system dynamics that would be hard won through empirical observations or brute-force numerical simulation. Indeed, as we'll see later, stability also has a large impact on how well we can forecast and what tools we choose to employ. For more complex simulation modeling, a working understanding of numerical methods, such as optimization and numerical integration, can be invaluable. However, while still important, both analytical and numerical methods infuse the modeling process with an excessive focus on determinism. Furthermore, all of these techniques combined are often insufficient for making specific predictions about any real population. Forecasting real systems requires a somewhat different set of skills, more focused on data and with a greater dependence on statistics and informatics. Finally, it requires that we shift the focus of our modeling to be less deterministic and to *think probabilistically* about both data and models. In this way of looking at prediction, we will make frequent use of probability distributions, not to add stochasticity, but to represent our imperfect knowledge of the world (that is, to capture uncertainties). Even if we believe the world is completely deterministic, we want to make probabilistic forecasts because our knowledge is always incomplete. The goal of the following example is to analyze a well-known and simple model, logistic growth, as a guide to thinking probabilistically about ecological forecasts.

## ▌2.2 EXAMPLE: THE LOGISTIC GROWTH MODEL

The logistic growth model is derived from the observation that in the real world, most populations seem relatively stable and are not changing exponentially, unlike the populations predicted by the exponential growth model. If our simplest possible model assumes that per-capita growth rate $r$ is constant, the logistic growth model is derived from the next simplest assumption, that per-capita growth is a straight line (figure 2.1)

$$\frac{1}{N}\frac{dN}{dt} = a + b \cdot N$$

This equation is equivalent to the more common parameterization

$$\frac{dN}{dt} = rN\left(1 - \frac{N}{K}\right)$$

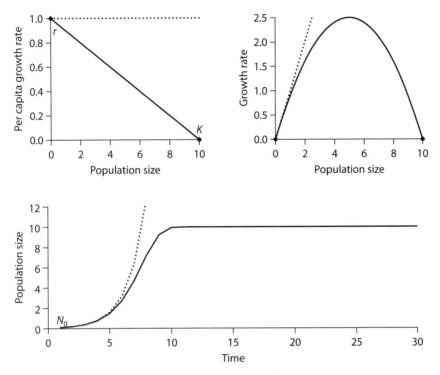

FIGURE 2.1. Logistic growth (solid line) with $r = 1$ and $K = 10$ compared to exponential growth with $r = 1$ (dotted line). The upper panels depict the relationships between population size and per capita growth rate (left) and population-level growth rate (right), with the two equilibria for the logistic shown as black diamonds. The lower panel depicts a discrete-time simulation of the logistic and exponential growth models starting from an initial population size $N_0$.

where $K$ is the carrying capacity, $a = r$, and $b = -r/K$. Note that there is no "ecology" in this derivation. Instead, a simple modification to the exponential growth model was introduced to produce a nonzero equilibrium. Yet the logistic growth model is probably the most well-known and influential population model in ecology, serving as the base from which many other models and theories have been, and continue to be, derived. In terms of traditional analyses, the logistic has two equilibria, $N = 0$ and $N = K$ (black diamonds in figure 2.1). If $r > 0$, then $N = K$ will be the stable equilibrium and $N = 0$ is unstable (and vice versa, if $r < 0$, then $N = K$ is unstable, and extinction, $N = 0$, is stable). Graphical analyses highlight that it is the sign of the growth rate, not the linearity of the growth curve, that matters—any other function with the same two intercepts will have very similar behavior.

Given the familiarity of the logistic model to most ecologists, I will use it as the basis for exploring ideas related to different sources of uncertainty that might affect our ability to make useful ecological forecasts. Because it is simpler to add sources of variability in discrete time, the discrete-time version of the logistic will be used in the following examples.

$$N_{t+1} = N_t + rN\left(1 - \frac{N}{K}\right)$$

## 2.3 ADDING SOURCES OF UNCERTAINTY

This section starts from the standard logistic model discussed earlier and progressively adds a number of sources of uncertainty and variability, such as *observation error, parameter uncertainty, initial condition uncertainty,* and *process variability.* Other important sources of error, which will be discussed but not simulated, include *model selection uncertainty, driver and scenario uncertainty,* and *numerical approximation error.* The goal of this section is to shift our focus and perspective to how a forecaster might approach the problem of making a prediction with well-quantified uncertainties. The terms uncertainty, variability, error, and stochasticity are used frequently in this chapter and throughout the literature. The key distinction in these terms are between *uncertainties,* which describe our ignorance about a process and in theory should decrease asymptotically with sample size, and sources of *variability,* which describe variation in the process itself that are not captured by a model. As sample size increases these can be better characterized but do not decrease.

### 2.3.1 Observation Error

One of the realities of science is that we can never observe a system without some noise or uncertainty in the observations themselves. Figure 2.2 shows an example of adding observation error to the logistic growth model—in this case, assuming that the error comes from a Normal distribution with a mean of zero and a standard deviation of 0.25 population units. The mean of zero implies that the observations are unbiased, while the standard deviation was chosen for illustrative purposes. In practice, the observation error would be estimated from data. In some cases, this is based on instrument calibration or similar controlled activities, but often there are other sources of observation error beyond instrument calibration (for example, sam-

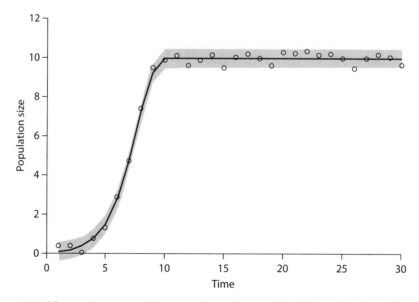

FIGURE 2.2. Observation error. Gray envelope shows a 95% CI, and points depict one stochastic realization of observation error.

pling error), and frequently observation error is inferred as part of the process of fitting the model to data. Indeed, traditional statistical estimation often assumes that observation error is the only source of error and fits models by attempting to minimize this residual error.

The key to understanding observation error is to realize that it does not affect the underlying process itself. The process is still deterministic, we're just observing it imperfectly. Indeed, most statistical models follow this assumption of conceiving residual error as being 100% observation error. It is also worth noting that observation error is always present and increasing the number of observations does not reduce the observation error—it does not disappear asymptotically like some other sources of error. However, it also does not affect a forecast made with this model, since it doesn't alter the underlying model dynamics. If observation error were the only source of error in figure 2.2 , then the forecast for the next year (time 31) would be that the population would be at carrying capacity (a population size of 10). Furthermore, if observation error was the only source of uncertainty, then this forecast would have zero uncertainty because observation error doesn't affect the system itself.

### 2.3.2 Parameter Uncertainty

Once we start discussing the idea of fitting the model to data we have to introduce the idea of parameter uncertainty—uncertainty about the true values of all the coefficients in a model. As with observation error, parameter uncertainty treats the underlying model as deterministic; however, we do not know the true values of the model's parameters. Parameter error in models is most familiar to ecologists as the standard error on a mean or the hourglass-shaped confidence interval (CI) that surrounds any linear regression. This hourglass shape arises from the combination of uncertainties in the intercept and the slope. The intercept moves the line up and down and by itself would produce a band-shaped confidence interval, while the slope changes the inclination of the line and by itself would produce a double-triangle shaped confidence interval.

Figure 2.3 illustrates the inclusion of parameter error into the logistic model, considering first $r$ and $K$ individually, and then their impact together. This model can be written as

$$N_{t+1} = N_t + rN\left(1 - \frac{N_t}{K}\right)$$

$$\begin{bmatrix} r \\ K \end{bmatrix} \sim N_2\left(\begin{bmatrix} r_0 \\ K_0 \end{bmatrix}, \Sigma_{param}\right)$$

where $\sim$ is read "is distributed as," $N_2$ refers to a two-dimensional Normal distribution, $r_0$ and $K_0$ are the means of these parameters, and $\Sigma_{param}$ is the $2 \times 2$ parameter error covariance matrix (see box 2.1). Not surprisingly, uncertainty in $r$ impacts only the growth phase, while uncertainty in $K$ has the largest impact in the asymptotic phase. Compared to other interval estimates we are familiar with, all three confidence intervals in figure 2.3 are considerably different from the classic hourglass-shaped regression confidence intervals. The bottom panel also depicts a number of stochastic realizations of model dynamics using different parameters drawn from their respective distributions. This highlights that, while the confidence interval reflects the distribution of parameter uncertainty, the underlying dynamics remain deterministic.

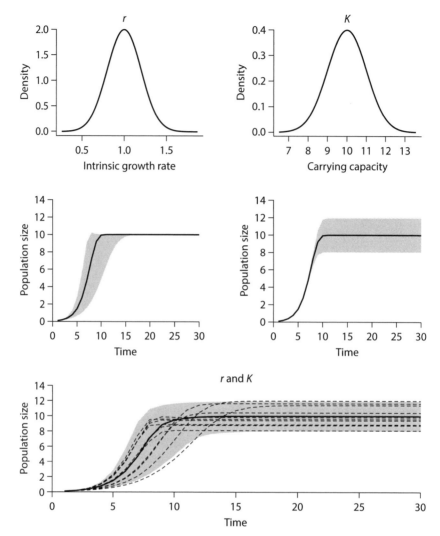

FIGURE 2.3. Parameter error. The parameters were assumed to be distributed $r \sim N(1.0, 0.2)$ and $K \sim N(10, 1.0)$. The left-hand column shows the distribution for $r$ and the median (solid line) and 95% confidence interval (gray). The right-hand column depicts the equivalent for $K$. The bottom panel depicts the combined effects of uncertainty in $r$ and $K$ as well as 20 realizations of the model dynamics (dashed lines) generated by drawing different parameter values from their distributions.

### 2.3.3 Initial Conditions

Just as there can be uncertainty in the model parameters, there can also be uncertainty in the initial conditions. The initial conditions specify the starting values of the system's *state variables*. State variables are those that give a snapshot of the system's properties at any single point in time. In the case of the logistic, the state variable is population size, and the initial condition would be the value of $N_0$, the initial population size. In other ecological models, state variables might include the composition

## Box 2.1. Covariance

*Covariance matrices* will come up frequently in this book, so it's worth taking a moment to understand what covariance is and how to interpret covariance matrices. Formally, the covariance between two variables $X_i$ and $X_j$ is

$$Cov(X_i, X_j) = \sum (X_i - E[X_i])(X_j - E[X_j])$$

where $E[X]$ is the expected value (mean). Variance is just the covariance between a variable and itself, and thus represents an important special case. Covariances are often not intuitive to read and evaluate by themselves, but any covariance matrix $\Sigma$ can be decomposed into a *diagonal matrix of standard deviations (D)*, one for each $X$, and a *matrix of correlation coefficients (R)*:

$$\Sigma = D \cdot R \cdot D$$

The diagonal of the correlation matrix will always be 1, since a variable is always perfectly correlated with itself. The off-diagonal values will be between −1 and 1, the same as any other correlation coefficient. Thus if two variables are independent, their covariance matrix will contain only the variances of each variable by itself on the diagonal and will be zero on the off-diagonal. However, we'll frequently encounter variables that are not independent, but rather have positive or negative correlations. Common examples of covariance matrices are in time-series and spatial analysis, where they are used to account for the nonindependence of measurements. Covariances are also very common in parameter estimates. For example, in a standard regression the slope and intercept typically have a negative covariance because any time you increase one (for example, the slope), you have to decrease the other (the intercept) to continue to make sure the line goes through the data. Finally, covariances are also critical to the "model as scaffold" idea introduced in chapter 1, where we use correlations among the variables being forecast by a model to allow us to constrain one variable based on observations of another, or to allow multiple types of data to be mutually informative.

of a community, the size or age structure of a population, or the pool sizes of biomass, water, or nutrients. By contrast, variables describing the changes of those states are typically not treated as state variables (recruitment, mortality, net primary productivity [NPP], and so on). Initial conditions often receive less attention in traditional ecological modeling because of the focus on equilibrium conditions. When initial conditions are discussed, it is often in terms of portions of parameter space rather than specific values. For example, in the Lotka-Volterra competition model, there are cases where the equilibrium reached depends *qualitatively* on the initial abundances of the different species; however, the *quantitative* location of the equilibria themselves are not a function of initial conditions.

A fundamental difference between theoretical modeling and real-world forecasting is the enormous amount of time and effort often put into understanding the current state of a system. One reason for this is that in forecasting we are often much

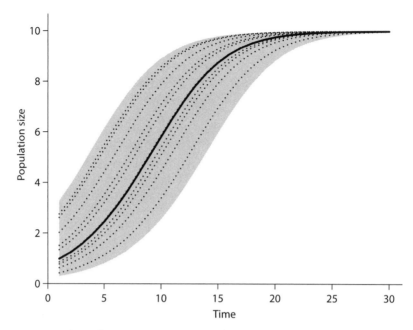

FIGURE 2.4. Initial conditions. Uncertainty in the initial population size was assumed to be lognormally distributed with mean 1 and standard deviation of 0.6. The logistic was modeled with $r = 0.3$ and $K = 10$. This figure depicts the median trajectory (solid black line), 95% interval (gray), and 10 stochastic realizations of different initial conditions.

more interested in the current trajectory of a system than its theoretical asymptote. Given the time lags in ecological dynamics and the importance of site history, a "transient" trajectory may persist for hundreds of years. In the case of American chestnut (*Castanea dentata* [Marsh.] Borkh), its post-glacial range expansion in the Holocene appears to have lagged its climate equilibrium by thousands of years (Davis 1983). Furthermore, in forecasting real systems, one frequently needs not just one initial condition but maps of the variable of interest, such as the current distribution and abundance of a species. More complex models often have multiple state variables, requiring the estimation of maps of multiple quantities. For example, even a simple terrestrial carbon cycle model will involve multiple plant biomass pools (for example, leaves, stem, roots), soil carbon pools (for example, labile, recalcitrant), and soil moisture pools (different depths) that need to be estimated. It is in estimating these initial conditions that the informatics challenges of ecological forecasting (chapters 3 and 4) often first become apparent.

Figure 2.4 shows the impact of uncertainty in initial conditions on the predictions of the logistic model. Here we are assuming that the initial population size follows a lognormal distribution with a standard deviation of $\tau_{IC}$

$$N(t = 0) \sim logN(N_0, \tau_{IC})$$

where the lognormal was chosen to reflect that the initial population size cannot be negative. In this case, the effects of the initial conditions eventually decay because the stable equilibrium predicted by the model is independent of initial conditions, but this is not always the case (box 2.2). The individual trajectories based on different

## Box 2.2. Chaos

Sir Robert May (May 1976) first discovered that the discrete time logistic model could produce chaotic dynamics over forty years ago. This has since become a classic example of how a simple model can generate complex dynamics. Most ecologists have seen the classic bifurcation diagram for the discrete logistic, which depicts how the system shifts from stable to periodic oscillations to chaotic as $r$ increases. Within the chaotic region the discrete logistic oscillates through time along what is a seemingly random trajectory, but which is completely deterministic (figure 2.5). Furthermore, the cycles are nonperiodic, which means that the population never returns to any size that is exactly the same as at any time in the past. In other words, the cycles never repeat. However, what defines chaotic systems is their *sensitivity to initial conditions*. Thus a population forecast with slightly perturbed initial condition will have a different trajectory from the original, with the two trajectories diverging more over time instead of converging to similar trajectories. Indeed, the initial perturbation doesn't just diverge slightly from the original, but the difference increases at an exponential rate. By contrast, in the nonchaotic case, two nearby trajectories

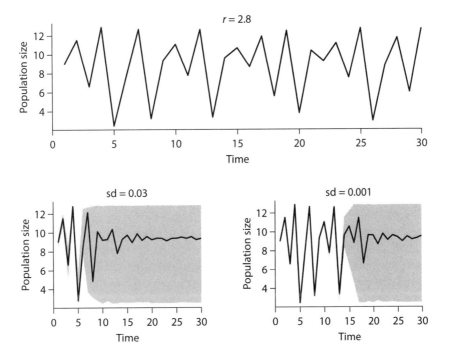

FIGURE 2.5. Sensitivity to initial conditions. The top panel depicts the chaotic trajectory of the discrete logistic with $r = 2.8$, $K = 10$, and $N_0 = 9$. The bottom panel depicts the median (solid line) and 95% interval (gray shading) for the same trajectory starting from a lognormal initial condition with standard deviations of 0.03 and 0.001, respectively.

*(Box 2.2 cont.)*

will converge at an exponential rate, unless the initial condition happens to be an unstable equilibrium point. The parameter describing the rate of this exponential convergence or divergence is called the *Lyapunov exponent*, and, as with $r$ in the exponential growth model, a positive value indicates divergence (chaos) and a negative value indicates convergence (stability). The interested reader is encouraged to read further about chaotic dynamics and their analysis in any of a number of excellent textbooks on theoretical ecology (Otto and Day 2007).

While most treatments of chaotic dynamics discuss the idea of a single small perturbation in the initial conditions, from the perspective of forecasting, the problem instead is that we have uncertainty in the initial condition, which we treat as a probability distribution. The lower two panels of figure 2.5 depict the same model for initial conditions as in section 2.3.3, $N(t = 0) \sim logN(N_0, \tau_{IC})$, but with two contrasting standard deviations ($\tau_{IC} = 0.03$ vs. $0.001$). As with the Lyapunov exponent, the width of the confidence interval (gray) grows exponentially in both cases. However, from a practical perspective, the shift in predictions from "fairly accurate" to "no better than random" appears to be a quite rapid. In effect, this transition defines the limit of predictability, which depends strongly on the initial uncertainty in the population size. It should be noted that, in our example, changing from an initial uncertainty of 0.03 to 0.001 is a 30× decrease in standard deviation but this increases the limit of predictability by less than 3×. Furthermore, because standard error declines in proportion to $1/\sqrt{n}$, where $n$ is sample size, a 30× reduction in uncertainty requires a roughly 900× increase in sample size. Even the less accurate of the two simulations has a high accuracy relative to most population estimates (<1% error).

While the discrete-time logistic can produce chaotic dynamics, it has arguably led to more "mathematical curiosities" than useful ecology. That said, we will revisit the issue of chaotic dynamics in chapter 15 when discussing measles outbreak dynamics (Bjørnstad et al. 2002). The possibility of chaos in such simple models has also raised important questions about the ability to distinguish chaos from stochasticity (Ellner and Turchin 1995). For ecological forecasting the more important question is how chaotic dynamics impact our ability to make forecasts. At first glance it would appear to make forecasting impossible, but in fact all of weather forecasting is done with chaotic models (Kalnay 2002), yet meteorology has been held up as a success story in forecasting (Silver 2012). Indeed, many of the tools and techniques ecologists are using for forecasting are borrowed from atmospheric science (section 2.5). However, these tools were primarily designed to constrain initial conditions for fundamentally chaotic problems (that is, problems dominated by initial condition uncertainty). One of the most basic questions in ecological forecasting is determining what the fundamental nature of the ecological forecasting problem is (section 2.5.3), as this is deeply connected to which sources of uncertainty will dominate ecological forecasts, which in turn affects not only the tools and techniques we use but also how successful we are likely to be in forecasting different processes.

initial conditions (dotted lines) show that, as with the parameter uncertainty, the underlying dynamics are deterministic, but there is uncertainty about which trajectory is the true one.

## 2.3.4 Process Error

Many ecologists are familiar with the idea of *stochastic models*, where random numbers are drawn that add noise to the dynamics of a model. Traditionally, this noise is used to represent two forms of variability, demographic stochasticity and environmental stochasticity (Lande 1993). Demographic stochasticity arises from the discrete nature of births and deaths. It can have large impacts on small populations, but quickly averages out for large populations. Environmental stochasticity is typically thought of as representing the unpredictable year-to-year variability associated with climate. This stochasticity is often modeled as proportional to population size, or as altering vital rates, so its effects tend to dominate larger populations. Of course, for large and widely distributed populations, different individuals are often not experiencing the same weather, which tends to have a stabilizing effect.

Both forms of stochasticity are examples of a more general phenomenon called *process error*. Process error encompasses *all of the sources of true variability in the underlying process that are not captured by the model*. At its essence, since all models are approximations of reality, it is the error associated with everything that's been left out. What is critical, however, is to realize that these errors *affect the underlying model dynamics*. They propagate forward in time, outward in space, and across any other unit being investigated (such as individuals, species, and so on). Furthermore, how this error is partitioned among these different sources can have large impacts on predictions. For example, given the same total variance, whether you have large plot-to-plot variability versus large year-to-year variability would produce very different predictions. The methods for partitioning uncertainty will be the subject of chapter 6. It is also important to realize that while process errors show up in the residual error in a traditional approach to modeling, traditional models are much more likely to treat that residual error as observation error than as true variability in the process. In that sense process error can also be thought of as the remaining residual error after observation error has been accounted for. However, it is important to realize that process error is not necessarily additive or independent but is often autocorrelated or reflects variability in model parameters, and both of these types of variability can have multiple spatial and temporal scales.

Figure 2.6 shows two examples of adding process error to the logistic growth model. In the first model the process error is assumed to be additive, which is similar to how residual error is typically specified except that, as noted earlier, this variability propagates through time because it impacts the population size the following year

$$N_{t+1} = N_t + rN_t\left(1 - \frac{N_t}{K}\right) + \varepsilon_t$$

$$\varepsilon_t \sim N(0, \tau_{add})$$

where $\tau_{add}$ is the additive process error standard deviation. The second model assumes that the process error occurs through variability in the model parameters themselves

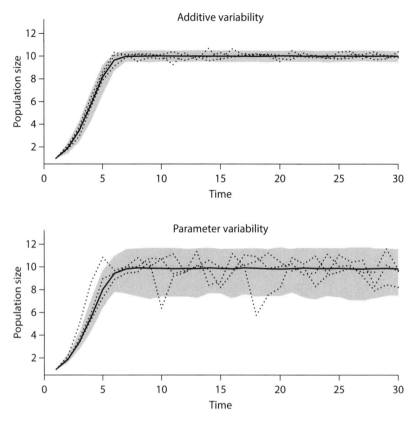

FIGURE 2.6. Process error. The top panel depicts adding $N(0,0.25)$ variability to the logistic growth model. The lower panel depicts allowing the model parameters to vary between time steps with $r \sim N(1.0,0.2)$ and $K \sim N(10,1.0)$. Both panels show the median (solid line), 95% interval (shaded), and three examples of stochastic realizations of the process (dotted lines).

$$N_{t+1} = N_t + r_t N_t \left(1 - \frac{N_t}{K_t}\right)$$

$$\begin{bmatrix} r_t \\ K_t \end{bmatrix} \sim N_2\left(\begin{bmatrix} r_0 \\ K_0 \end{bmatrix}, \Sigma_{process}\right)$$

where the subscripts on $r_t$ and $K_t$ indicate that they vary through time, and $\Sigma_{process}$ is the process error covariance matrix for the parameters $r$ and $K$, which have means $r_0$ and $K_0$. Finally, it is also sensible in many circumstances to include multiple sources of process error, so long as they are not redundant. For example, in a regression model, additive process error and process error in the intercept would be mathematically identical, so we would not want to include both.

### 2.3.5 Other Sources of Uncertainty

The sources of uncertainty discussed earlier are not the only ones relevant to forecasting, just the ones most easily demonstrated with the logistic growth model. Another common source of error in ecological forecasting is *model choice*. Since most ecological models are not based on physical or chemical laws, there is often uncertainty surrounding the choice of model, such as the choice between logistic and exponential growth. In general, increasing the complexity of a model increases parameter error but reduces process error. The classic model selection problem is how to choose the model that best balances these two sources of uncertainty (Gelfand and Ghosh 1998). However, model selection isn't guaranteed to give you the "correct" model, since the choice of "best" model is very much a function of the amount of data at hand and there is no guarantee that any of the models being considered are correct. In addition, model selection is far easier to apply to simple statistical models than to complex computer models. While model structure is not a continuous random variable, like most of the other sources discussed, it is still possible and generally beneficial to work with a set of models that make different structural assumptions. This set is often referred to as an *ensemble* (chapter 11). While some forecasting is done with unweighted ensembles (all models get an equal vote), there are other methods for model averaging that allow probabilities to be assigned to different models (Hoeting et al. 1999), essentially treating model structure like a discrete random variable (chapter 14).

In addition to predicting outputs, many ecological models require a range of inputs, such as weather, topography, soils, and so on. Uncertainty in these inputs is often referred to as *driver uncertainty* or *boundary condition uncertainty*. While this uncertainty is technically observation error, these uncertainties do propagate through the models because they apply to inputs rather than outputs. Closely related to driver uncertainty is *scenario uncertainty*, which typically refers to the uncertainty associated with the choice between different discrete scenarios for how the future may play out. Finally, there are *numerical approximation errors* any time you rely on a computer to perform a calculation. In many cases these errors are trivially small compared to other sources of error in ecological models, but this is not always the case. For the most part I will not touch on the computer science and applied math surrounding numerical approximation, as there are already many excellent texts on this subject (Press et al. 2007).

## 2.4 THINKING PROBABILISTICALLY

In the traditional modeler's view, there are three parameters in the logistic model, $r$, $K$, and $N_0$, with the first two receiving the lion's share of attention in most discussions (figure 2.7). If we instead think about putting together the sources of uncertainty discussed earlier to capture the variability in any real ecological system, what additional parameters need to be considered? To start there's observation error (section 2.3.1), which we earlier assumed is unbiased and Normally distributed, and so can be represented with one variance, $\tau_{obs}$. Next, for parameter uncertainty (section 2.3.2), we're interested in the variances and covariance of two parameters, $r$ and $K$, which we'll represent as the $2 \times 2$ covariance matrix $\Sigma_{param}$. In terms of process error (section 2.3.4), there's the additive noise, $\tau_{add}$, as well as the variability in the parameters themselves, which we can represent with another $2 \times 2$ covariance matrix, $\Sigma_{process}$. In addition, uncertainty in the initial conditions (section 2.3.3) can be represented as $\tau_{IC}$. All together, for a three-parameter model, there are nine additional variances and covariances to be estimated to describe the uncertainty in both the data

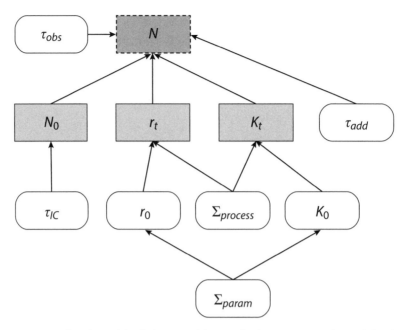

FIGURE 2.7. Graphical model of the variables in the logistic growth model. The traditional approach focuses on the prediction of the state variable, $N$ (dashed dark gray box), at any point in time given the initial conditions, $N_0$; the intrinsic growth rate, $r$; and the carrying capacity, $K$ (solid light gray boxes; arrows indicate that one parameter affects another). Accounting for uncertainty adds observation error ($\tau_{obs}$), parameter error ($\Sigma_{param}$), process error ($\tau_{add}$, $\Sigma_{process}$), and initial condition uncertainty ($\tau_{IC}$). Process error includes the possibility that the model parameters vary through time ($r_t$, $K_t$); thus $r_0$ and $K_0$ describe their mean values.

and the model. The larger number of variances is indicative of the shift in emphasis and effort that occurs when moving from modeling to forecasting. This example is hardly exhaustive, as many of these parameters will vary systematically (for example, in response to environmental heterogeneity in space or climate variability in time) or may vary randomly across a wide range of spatial, temporal, or taxonomic scales (such as differences among subpopulations). For example, one may need to estimate a whole map of initial conditions describing where a species is present and how abundant it is at each location. *Furthermore, this description of the sources of variability and uncertainty does not tell us which of these parameters and variances are most important (that is, which have the greatest impact on forecast uncertainty).* Indeed, that question cannot be answered a priori and will depend on the relative uncertainties and sensitivities of different variables as well as the time frame considered. For example, in nonchaotic systems initial condition uncertainties often dominate short-term predictions, while process, parameter, and scenario uncertainties are frequently more important for longer-term projections (Hawkins and Sutton 2009).

Compared to traditional modeling (section 2.1), the most striking difference in this example is that *everything in the model is described in terms of probability distributions.* Rather than thinking about the model parameters as constants, we are acknowledging uncertainties in all aspects of the model forecast. These uncertainties are included by treating parameters as *random variables*, which formally means that they do not take on a single value but that they can take on a range of different values described by a probability distribution. This is a fundamentally different way of thinking from just adding stochastic noise onto a deterministic model, as it puts probability at the center of the modeling process. Modeling itself becomes a process of combining and transforming probabilities, and probability distributions become the core building blocks in model development. Philosophically, treating parameters as random variables doesn't necessarily mean that the process itself is stochastic. It is a statement about our imperfect understanding of the process, not a statement about the nature of the process itself. Indeed, even from a completely deterministic world view, it makes sense to accommodate our imperfect knowledge of the world using probability distributions. Overall, one of the most important concepts of this book is the need to think about all aspects of models and data from a probability theory perspective.

## 2.5 PREDICTABILITY

Before diving into the nuts-and-bolts of data-model fusion and ecoinformatics, it is important to think about the predictability of ecological systems and the nature of the *ecological forecasting problem*. Even outside of ecology, forecasting is fraught with many examples of failure. As Niels Bohr famously joked, "Prediction is very difficult, especially if it's about the future." In casting a wide view across many disciplines, popular writer and political forecaster Nate Silver identified the traits of successful versus unsuccessful forecasts (Silver 2012). (Note: Armstrong [2001] offers a more detailed 89-point checklist of forecasting recommendations.) At the top of this list is the importance of thinking probabilistically and communicating uncertainty (section 2.4), and he goes as far as to state that "uncertainty is an essential and non-negotiable part of a forecast." Silver also stresses the importance of not being locked into a single hypothesis, model, or approach, but to think self-critically and to entertain an ensemble of multiple competing hypotheses or models, whether internally or across different research groups. Similarly, he notes the importance of updating forecasts as new information becomes available, and to avoid the temptation of overfitting models to data, as our natural inclinations to find patterns and defend our positions can both increase forecast errors. All of these suggestions are not just recommendations for good forecasts, they are fundamentally part of doing good science.

Some of the greatest failures in forecasting are failures of imagination—that is to say, failing to anticipate low-probability events that may have substantial impacts on the system being forecast. As former US Secretary of Defense Donald Rumsfeld famously said (US Department of Defense 2002):

> There are known knowns; there are things we know that we know.
> We also know there are known unknowns; that is to say, there are things that we now know we don't know.
> But there are also unknown unknowns—the ones we don't know we don't know.
> And if one looks throughout the history of our country and other free countries, it is the latter category that tend to be the difficult ones.

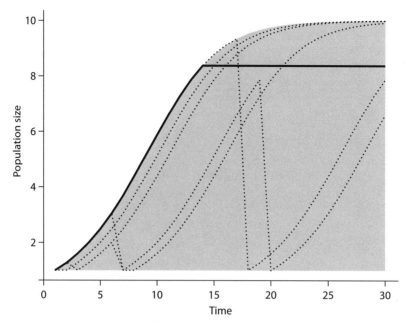

FIGURE 2.8. Adding a stochastic 5% per year probability of disturbance to the logistic growth model ($r = 0.3$, $K = 10$) causes the confidence interval (gray shading) to encompass all possible states. The solid black line shows the median prediction, while the dotted lines show examples of individual ensemble members.

In the context of section 2.4, the known unknowns are things that we put a probability distribution on. The unknown unknowns are the things we fail to anticipate completely—what Nassim Taleb calls "black swans" (Taleb 2007). The name "black swan" comes from the unquestioned assumption that all swans are white because all swans ever observed had only been white—that is, until European explorers discovered black swans in Australia. As an example, if we introduce a small probability of disturbance to our earlier logistic growth model, the confidence interval on our forecast blows up (figure 2.8). Failing to anticipate this probability of disturbance does not actually have a huge impact on the mean but results in enormously overconfident results. Interestingly, if we consider the disturbance process at a large scale, then the impacts of fine-scale disturbance will average out and the system remains predictable, provided that such disturbances are *independent* of one another. If, on the other hand, disturbances are highly correlated with one other (for example, a hurricane), then the large-scale confidence interval is essentially the same as the fine-scale one. Thus, the usefulness of such a forecast hinges on what may appear to be a small assumption about the covariance structure, and one that may be very hard to estimate from the data. The only real solution is to integrate over that uncertainty and to understand and acknowledge what sources of uncertainty are limiting predictability.

### 2.5.1 Barriers to Forecasting

The disturbance example highlights that there may be *inherent barriers* to certain types of ecological forecasts. While they come in multiple forms, in essence these are uncertainties that are not substantially reduced by the addition of any practical

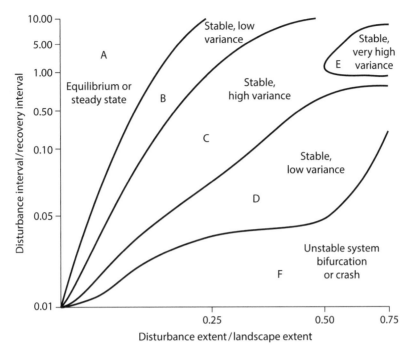

FIGURE 2.9. State-space of ecosystem predictability in relationship to disturbance extent relative to landscape extent (*x*-axis) and disturbance return interval relative to recovery interval (*y*-axis). From Turner et al. (1993).

amount of data. Stochastic high-impact events, such as local disturbances, are one such example, but not all disturbance regimes fit this category, as many occur at a frequency, size, or intensity that is much more predictable at a landscape scale (Turner et al. 1993) (figure 2.9). As discussed earlier in box 2.2, even deterministic systems can be challenging to forecast if they demonstrate a high sensitivity to initial conditions. Chaotic systems often possess a limit to predictability in time, beyond which the trajectory of the system is indistinguishable from its long-term summary statistics (for example, mean, variability). Another potential barrier is the phenomenon of *computational irreducibility*, whereby a complex system cannot be approximated by a simple model (Beckage et al. 2011). Computationally irreducible systems cannot be modeled except by an explicit representation of detailed processes and their dynamics through time, which is often unfeasible. Beckage argues that such properties may be common for ecological phenomena, such as species niche spaces, and limit the predictability of important global change processes, such as climate change and exotic species invasion.

Finally, predictability is fundamentally reduced by *failures to understand* the system being forecast. For example, economic time-series models are often prone to overparameterization and limited forecast accuracy because they are primarily based on observed patterns. This is also a common problem in many areas of ecological modeling, where there is a tendency to solve model failures by adding things to the model. This increase in model complexity may not always be accompanied by the availability of data, and thus may lead to a reduction in parsimony. By contrast, the mechanistic models of atmospheric physics used for weather forecasting are able to accommodate much greater complexity because they represent physical processes. There is concern that the rise of "big data" and data mining techniques will fail to improve forecasts if they are not coupled to clear hypotheses and an understanding of the underlying processes.

### 2.5.2 Weather Forecasting

Weather forecasting is often held up as an example of successful forecasting in the environmental sciences. Indeed, despite weather forecasters bearing the brunt of numerous jokes, numerical weather forecasts have steadily and consistently improved in skill over the last 60 years (figure 2.10). However, if we focus on the start of this graph, one of the important messages is that the much younger discipline of ecological forecasting should expect to be bad in the beginning. But initially poor forecasts should not dissuade us from forecasting, so long as the uncertainties are accurately reported, as we will not get better until we get on the learning curve. Ecologists can no longer wait to start making forecasts, and it is unrealistic to postpone forecasting until we have perfect knowledge of the systems we are studying. Furthermore, forecasting itself will provide an efficient means for improving our understanding of the systems we study. Part of the success of weather forecasters is that every day they perform a global experiment and get to compare their predictions to reality. More generally, weather forecasting has advanced through steady improvements in models, data quantity and quality, computation, and the process of bringing data and models together. Furthermore, quantitative forecasts are not taken as truth, but are interpreted from the perspective of local human knowledge and experience by individuals that know the flaws in the computer models.

Many of the tools for model-data fusion developed in the atmospheric sciences are being transferred to other environmental science disciplines, including ecology, and

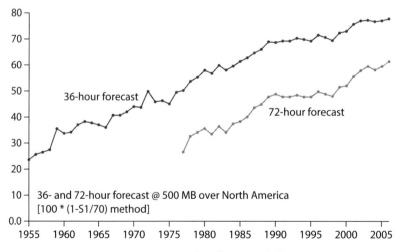

FIGURE 2.10. Improvements in NCEP weather forecast skill score (y-axis, 100 = perfect score, 0 = random) over a 50-year period. S1 skill combines both the model-data mismatch and the gradient in both. Adapted from http://celebrating200years.noaa.gov/foundations/numerical_wx_pred/S1Chart06.html.

many of these will be covered in this text (chapters 13 and 14). In weather forecasting there clearly are a wide range of uncertainties in an atmospheric model. On the one hand, while there are definitely empirical parameterizations in weather models, compared to ecological forecasts these models generally rely more on physical laws, which reduces parameter uncertainty and process error. On the other hand, the atmosphere is highly chaotic, which makes these models very sensitive to initial conditions. The model-data fusion problem for weather forecasting has thus traditionally been to better and better characterize the initial conditions and their uncertainty so that the deterministic model can project into the future. By the characteristics considered earlier, the atmosphere is fundamentally unpredictable; however, it is the archetype of a system that has become forecastable. It is an open question how many of these tools can be used as-is by ecologists and which will need to be adapted to the different nature of the forecasting problem in ecology.

### 2.5.3 The Ecological Forecasting Problem

As noted earlier, there are important differences between the nature of the forecasting problem in meteorology versus ecology. To gain a better understanding of this and how the different sources of uncertainty we've discussed impact ecological forecasts, let's step back and see what we can learn from a simple first-principles approach. Consider the general case of a dynamic model predicting the state of the system, $Y$, in the future as a function of the *current* state of the system, some covariates (or drivers) $X$, and parameters $\theta$. This general model doesn't describe all possible types of ecological forecasting problems, but it does capture major types of models, used across all ecological subdisciplines, to model populations and communities as well as ecosystem pools and fluxes. To elaborate a bit more on this model, let's also assume that it contains both additive process error, $\varepsilon$, and variability in the model's parameters, which we'll represent as deviations, $\alpha$, around some overall mean $\bar{\theta}$. Put together this gives us the following model:

$$Y_{t+1} = f(Y_t, X_t | \bar{\theta} + \alpha) + \varepsilon$$

which is just the general case of our earlier logistic model.

Given this general model, let's look at how the different factors involved affect our ability to make forecasts. To do so we will focus on the uncertainty in the forecast, which we'll summarize in terms of its variance, $Var[Y_{t+1}]$. Using techniques introduced in chapter 11, this variance is approximately

$$Var[Y_{t+1}] \approx \underbrace{\left(\frac{\partial f}{\partial Y}\right)^2}_{\substack{stability}} \underbrace{Var[Y_t]}_{\substack{IC \\ uncert}} + \underbrace{\left(\frac{\partial f}{\partial X}\right)^2}_{\substack{driver \\ sens}} \underbrace{Var[X]}_{\substack{driver \\ uncert}} + \underbrace{\left(\frac{\partial f}{\partial \theta}\right)^2}_{\substack{param \\ sens}} \left(\underbrace{Var[\bar{\theta}]}_{\substack{param \\ uncert}} + \underbrace{Var[\alpha]}_{\substack{param \\ variability}}\right) + \underbrace{Var[\varepsilon]}_{\substack{process \\ error}} \quad (2.1)$$

This equation may be long, but it follows a simple pattern—a sensitivity times a variance—repeated for each of the factors we're considering. In the following text we will break down some of the expectations for each of these factors in general terms, but the power of this equation is that for any particular application it gives a *quantitative* way of breaking down the overall forecast into its components and comparing their relative size and importance.

The first term in equation 2.1 describes how the initial condition uncertainty propagates through time, which is controlled by the sensitivity of the model to its current state. Circling back to the beginning of this chapter where I introduced the exponential, this particular sensitivity goes by another, more familiar name: *internal stability*. Indeed, this is the exact same stability used in classic ecological theory—so classic theory definitely plays a part in understanding ecological predictability! Furthermore, we can rely on the same criteria as classic theory to understand these dynamics. Because equation 2.1 is a recursive model ($Var[Y_{t+1}]$ is a function of $Var[Y_t]$), if $|\partial f/dY| > 1$, then the initial condition uncertainty will grow exponentially and dominate all other terms. This is precisely what happens in chaotic systems and why we say that weather forecasting, which deals with a chaotic system, is inherently an initial condition problem. By contrast, if $|\partial f/dY| < 1$, which is to say that the internal dynamics of the system have stabilizing feedbacks, then the initial condition uncertainty will decay exponentially with time. That's not to say that this uncertainty will be small—it may very well dominate short-term forecasts—but it does suggest that the other sources of uncertainty will come to overshadow initial conditions. As discussed earlier (box 2.2), there are definitely ecological systems that appear to be chaotic, such as epidemic disease—in these cases the first-principles approach makes a strong prediction about the nature of the ecological forecasting problem. However, searches for chaos more broadly across ecology have met with mixed results, and there are numerous examples of stabilizing feedbacks in ecology, so my personal working hypothesis is that for most ecological forecasting problems we need to focus on the other terms. This is admittedly a hypothesis and one I'd love to see tested. The final important lesson we learn from the first term of equation 2.1 is that it is the *only* term that has this property of providing a recursive feedback. In other words, it is the only term that will grow or shrink exponentially. All other terms in equation 2.1 will instead respond linearly over time. Not only will they respond linearly, but, because both the variances and the squared sensitivities are always positive, they will all *grow* linearly. This means that, in general, *the uncertainty in a forecast will increase into the future*. However, for $|\partial f/dY| < 1$, it can also be shown that this uncertainty will eventually reach a steady state, where internal feedbacks are sufficiently stabilizing to offset all other terms in equation 2.1.

If the first term in equation 2.1 described a system's *endogenous (internal) stability*, then the second term describes its *exogenous (external) sensitivity*—its sensitivity to external forcings. While the magnitude of this term will be problem specific, it is possible to make some general statements about our expectations. First, let's think about Var[X] itself, the variability in the model's covariates or drivers. Equation 2.1 tells us that systems that are sensitive to unpredictable drivers will themselves be unpredictable. By contrast, higher predictability can arise either from an insensitivity to environmental variability (for example, a resilient life-history stage, such as a seed bank, that provides a storage effect) or sensitivity to drivers that are themselves highly predictable (for example, synchrony to diurnal, tidal, or annual cycles). The other generality is that making a forecast of Y that depends on X requires that we be able to forecast X as well. Since, as noted previously, forecasts will tend to increase in uncertainty through time, this implies that unless X has strongly stabilizing feedbacks, an increase in Var[X] through time will cause an increase in the importance of the exogenous uncertainty through time. That said, some drivers will obviously increase in uncertainty faster than others (for example, weather) and some may be so slow to change as to be essentially constant for many forecasts (for example, topography, soils). Another important lesson from this is that *we might use different covariates for forecasting than we use for explaining the same process.* For example, when compared to past data, $X_1$ might be a better predictor of $Y$ than $X_2$. However, the future uncertainty in $X_1$, $Var[X_1]$, might be larger than that for $X_2$, in which case $X_2$ might be a better variable to choose when forecasting. How much better does $X_1$ have to be, or how much more certain does $X_2$ need to be, to know which to use in a forecast? Thankfully, equation 2.1 gives us a precise prediction on this matter: if $(\partial f/\partial X_1)^2 Var[X_1] + Var[\varepsilon_1] < (\partial f/\partial X_2)^2 Var[X_2] + Var[\varepsilon_2]$, then $X_1$ will produce a lower uncertainty forecast than $X_2$. The other important lesson from this analysis is that *experimental design for forecasting is often different from that for hypothesis testing.* In the standard hypothesis testing framework, the question is often about whether $X$ affects $Y$, which is frequently tested using an ANOVA design—impose discrete treatments of $X$ and ask if the $Y$'s are different. By contrast, in forecasting the central question isn't whether $X$ affects $Y$ but *how much*—what is the slope $|\partial Y/\partial X|$, or more generally, what is the shape of the relationship between $X$ and $Y$. This question requires a regression design—vary $X$ across a continuum of values spanning the expected range of variability.

The third term in equation 2.1 addresses the effects of *parameter uncertainty* and variability. We will return to these topics of estimating (chapter 5), partitioning (chapter 6), and analyzing (chapter 11) these uncertainties throughout this book, but here I want to think about their impacts in general terms. First and foremost, as noted earlier, is to recognize the difference between these two components, uncertainty and variability, as they have very different impacts on forecasts. Parameter uncertainty is the core focus of most statistical training: How well do we know the mean of $\theta$? Classic statistical theory tells us the answer to this question is largely a question of sample size. For problems where data are sparse, parameter uncertainty will be large and can often dominate ecological forecasts—this will be the case for chronically data-limited problems such as newly emerging diseases and invasive species. However, classical theory also tells us that $Var[\theta]$ will asymptotically decline to 0 as the amount of data increases, typically in proportion to $1/\sqrt{n}$. Therefore, provided any forecasting problem takes an iterative approach of incorporating new data as it becomes available, parameter uncertainty will tend to decline with time. Furthermore, equation 2.1 provides us with important guidance for how to reduce that uncertainty most efficiently (a topic we'll explore in much greater detail in chapter 11). Specifically, $(\partial f/\partial \theta)^2 Var[\bar{\theta}]$ implies that it is not parameter uncertainty or parameter sensitivity alone that matters, but that they matter equally. A parameter in a model can be important for either reason, while it is also possible for an uncertain parameter to be unimportant if it's also insensitive, or a sensitive parameter to be unimportant if it's well-constrained. In a model with multiple parameters, $(\partial f/\partial \theta)^2 Var[\bar{\theta}]$ gives us a way of ranking their importance based on their relative contribution to forecast uncertainty and then focusing research on constraining the most important terms first— what I called the model-data loop in chapter 1.

In contrast to parameter uncertainty, *parameter variability* does not converge asymptotically, it represents variability in the underlying ecological process itself. Thus, like we did in the earlier logistic example, we will consider this term as another manifestation of our final term in equation 2.1, *process error*. Process error in general is capturing all the variability not explained by the model. Part of this variability can be attributed to ecological factors that are *irreducibly stochastic*, such as reproduction, mortality, dispersal, and disturbance. I consider these sources of stochasticity to be irreducible because, like a coin flip, the amount of knowledge about the physical system and computation required to approach the problem deterministically is so incredibly vast and detailed, and on a spatial and temporal scale so divorced from the process of interest, that there's no conceivable way to distinguish the problem from true randomness, nor any reason to do so. The other major part of process error arises due to model *structural uncertainty*. Part of this error is because the equations we use in ecological models are rarely deterministic physical laws, and instead are often based on empirical correlation and variations on a few simple equations for population growth and mass and energy flux. The other part of this is that even when the equations themselves are reasonable, because the parameters in those models reflect biology rather than physical constants, there can be variability in the parameters themselves. This parameter variability reflects the reality that ecological processes are *heterogeneous* in space, time, and phylogeny across a wide range of scales. Accounting for parameter variability is an important tool for using measurements from one unit of measure (for example, population, watershed, or individual; year; species) to make inferences about another in a way that accounts for the larger uncertainty when doing so. Unlike the earlier logistic example, where the only form of process error we looked at was year-to-year variability at the population level, the random differences among other ecologically meaningful units—such as individuals, locations, and species—are often persistent, meaning that the differences are either permanent or slowly changing relative to the process of interest.

Statistically, we treat this parameter variability using random or hierarchical effects (chapter 6). Indeed, for the purposes of ecological forecasting, it is often *more pressing to quantify variability than it is to explain it*, since we want to avoid falsely overconfident forecasts. That said, as we learn more about a system, often we can chip away at this process uncertainty and attribute more and more of the observed variability to the deterministic components of our models (statistically speaking, these are "fixed effects"). Doing so requires the ability to quantify uncertainty in terms of probability distributions (chapter 5), the ability to partition that uncertainty into multiple sources (chapter 6), analyses to identify the dominant sources of uncertainty (chapter 11), and the ability to bring together multiple data sources (chapter 9) that highlight different aspects of a problem (for example, observations, experiments, physiological constraints). Taken as a whole, it is not obvious whether process error and parameter variability will increase, decrease, or stay the same over time. In terms of an individual forecast, if parameters are variable in space and time, they should tend to become more uncertain further out into the future. When viewed as an iterative process, the more measurements we make, the more we can refine our models and move uncertainty from process error into the dynamics. Nonetheless, there will always be something we didn't measure and didn't include in the model, so there will always be some portion of uncertainty that remains process error. Likewise, the presence of stochasticity, and in particular the risk of low-probability but high-impact events, will always continue to introduce uncertainty into ecological forecasts.

Overall, this first-principles perspective on the ecological forecasting problem (equation 2.1) provides a useful general framework. First, it provides a qualitative structure for thinking about the important factors that determine predictability that provides a number of explicit predictions and recommendations. Second, it provides a quantitative framework for comparing the absolute and relative magnitudes of different contributions both within a given problem and across different ecological processes. However, one thing that is currently lacking is a systematic, quantitative understanding of the relative contributions of these factors, and both how and why they vary from problem to problem. Such insight would get at the heart of many key questions in ecology, helping us better understand what drives ecological dynamics and allowing us to look for patterns and generalities that span all subdisciplines. Given limited time and resources, such an understanding would likewise be eminently practical, as it would allow us to better target resources at understanding the dominant sources of variability for a particular class of problem, rather than taking an "all of the above" approach where we have to tackle all sources of uncertainty equally. In the remainder of this book we will explore how this can be done and the current state of forecasting in ecology.

## 2.6. KEY CONCEPTS

1. Forecasting focuses modeling on real-world systems.
2. Think probabilistically! Uncertainties in models can be described by probability distribution functions.
3. There are a wide range of sources of uncertainty in ecological forecasts, such as observation error, parameter error, initial condition error, process error, model structure, driver uncertainty, scenario choice, and numerical approximation. Some of these sources of uncertainty propagate into forecasts while others do not.
4. It is important to distinguish uncertainties, which are a statement about our ignorance and decline asymptotically with additional data, from sources of variability, which describe the variation in the system itself.
5. Sensitivities to initial conditions, stochasticity, computational irreducibility, and failures in understanding impose limits to predictability.
6. A first-principles approach to the ecological forecasting problem provides a structure for partitioning and understanding the major uncertainties facing forecasts and linking those to the endogenous, exogenous, and parameter sensitivities of the model.

## 2.7 HANDS-ON ACTIVITIES

https://github.com/EcoForecast/EF_Activities/blob/master/Exercise_02_Logistic.Rmd

- Simulating the discrete logistic
- Probability distributions in *R*
- Monte Carlo simulation
- Parameter error
- Initial condition error

# 3

## Data, Large and Small

*Synopsis: Data must be organized, described, preserved, and made accessible if they are ever to be assimilated. This chapter introduces the tools and concepts needed to deal with modern data management, from the scale of individual researchers preserving their own data to the synthesis of large, distributed, and online data sources.*

### 3.1 THE DATA CYCLE AND BEST PRACTICES

Data are at the heart of ecological forecasting. As the volume and diversity of data that ecologists collect continues to increase, the skills and tools needed to manage such data are becoming more important. *Data management* can be defined as "the development, execution and supervision of plans, policies, programs and practices that control, protect, deliver and enhance the value of data and information assets" (DAMA International 2010). Ecologists have often shied away from data management under the false preconception that it is not work whose rewards are as tangible or immediate as time spent in the field or lab, doing analyses, and writing papers or reports. Nonetheless, a deeper appreciation of the value of good data management is emerging among many ecologists. Unfortunately, this often emerges out of regrets about how our past self failed to manage data, when everything was still fresh in our memory, or out of frustrations with the poorly managed data of others. Indeed, our present self tends to greatly overestimate how much we'll remember about what we're doing right now and how long we'll remember it for. And our past self tends not to answer our desperate e-mails (figure 3.1).

Adhering to best practices for data management first and foremost benefits ourselves—while it may cost time in the short run, it will ultimately save time and frustration in the long run. Beyond this, data citations and digital object identifiers (DOIs) are increasingly providing direct acknowledgement and reward for sharing data. More and more, data are seeing a second life in analyses that were never envisioned by the original data collectors, often to address questions or problems that didn't exist when the data was collected (Jones et al. 2006). Despite this, less than 1% of the data in ecology are currently recoverable post-publication (Reichman et al. 2011).

In addition to benefiting ourselves and our collaborators, more and more, there are ethical and legal demands for good data management. First and foremost, good data management and data sharing are key parts of transparency and repeatability, which are basic goals for all science and increasingly a requirement for science to inform decision making. The perspective that "Data not readily made available should

FIGURE 3.1. Documentation is written first and foremost for our future selves. Source: http://xkcd.com/1421/.

not be given credit in a debate" is becoming increasingly common and is fundamentally consistent with scientific skepticism (Schimel 2016). While most readers won't go through the effort of repeating the analyses in a published paper, that they should be able to is an important scientific principle. Furthermore, as the tools for leveraging prior research (chapter 5), performing synthesis (chapter 9), and data assimilation (chapters 13 and 14) become increasingly common, the frequency of data reuse is increasing.

For most of us the data we're collecting or working with were paid for by others, and thus we may feel a moral obligation to make that data available. Data collected using government funds belong to the public in a form that is accessible and well preserved. Increasingly, this obligation is becoming a legal one in many countries rather than just an ethical one (Beardsley 2014). That said, ethical requirements for data management can be very complex, as many data sources (for example, human subjects) come with legal restrictions, while in other cases there may be ethical reasons not to share data (for example, locations of endangered species populations). Indeed, cases with legal or ethical restrictions on data sharing are usually ones where adhering to best practices is most important. This highlights a final reason to adhere to best practices—whether in policy, management, or basic science, having transparent data management protects against accusations of misconduct.

When done well, data management is not an activity done at the end of a project but something that occurs through all stages of research. The DataOne project describes this as the "Date Life Cycle" (figure 3.2) (Michener and Jones 2012; Strasser et al. 2012), though similar ideas have been discussed by others (Jones et al. 2006; Reichman et al. 2011; Hampton et al. 2013, 2014; White et al. 2013; Goodman et al. 2014). Data management occurs primarily at two scales in ecological forecasting.

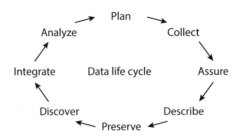

FIGURE 3.2. Data life cycle (Strasser et al. 2012). Image courtesy of DataONE.org.

First, it occurs at the data generation scale, when raw measurements are collected, checked, described, and preserved. Second, it occurs during synthesis and forecasting, when data are discovered by others, integrated, and analyzed. Activities at these scales may be carried out by the same team from start to finish, but increasingly synthesis and forecasting involve the integration of data from many teams, often without any direct communication other than public data documentation.

Overall, while the details of best practices in data management will not remain static as technology and scientific culture evolve, there are some key overarching principles that are likely to stand the test of time. The first is that data management should make working with data more *automated*. A key to this is that data be both machine and human readable and adhere to community standards where they exist. Second, data management needs to be *documented*, providing both metadata and information on data provenance so that any processing is traceable and repeatable. Finally, data must be *recoverable*, which means that the data must be archived in their original raw format, and in formats and locations that are going to stand the test of time.

### 3.1.1 Plan

Data management begins before any data are collected by constructing a data management plan that lays out how the other steps will be completed and by whom. Such plans are now required by many funding agencies as part of grant proposals. The goal of a good plan is to anticipate data challenges before they happen, rather than respond to them after the fact, and thus can be viewed as an integral part of experimental design to ensure data are usable. Time invested in planning for all other stages in the data life cycle will make the difference between whether data management becomes a simple and routine habit throughout a project versus an onerous and time-consuming task post hoc.

### 3.1.2 Collect

The goal of data management at the time of data collection and entry is not just to ensure that everything gets recorded, but to record data in a way that streamlines the whole data cycle and greatly reduces later headaches. Prior to collecting data, define all the variables being used, their units, and any codes to be used (for example, species abbreviations). This includes codes for missing data, as why data are missing should not be ambiguous (this has a surprisingly large impact on inference, as we'll see in chapter 6). There is no universal agreement on the choice of a missing data value, though the current suggestions (White et al. 2013) are NA, NULL, or a blank cell. A general theme in data management, analysis, and modeling is that it is better for an analysis to fail clearly at the source of the error than to silently continue on producing incorrect results, as can occur when using 0 or –999, which could easily be interpreted as data.

For tabular data, rather than using unstructured notebooks, use premade standardized tally sheets in the field and lab that contain variable names in a header row. Make sure that only one type of data is recorded on each sheet and that each observation is in its own row (though an observation may consist of multiple different measurements made on the same individual or sample). To make data easier to search, organize, and quality control, data should also be "atomized," with only one piece of information in a given record (for example, street, city, state, and zip code should be separate entries, rather than one entry for address). Where possible, data should be stored in standard, nonproprietary file formats (for example, csv rather than xls), as data that depend on a specific version of software, or software that must be purchased, quickly become inaccessible. Files should be stored with meaningful and unique file names that include a version number or revision date so that it is always clear what file version was used in any particular analysis. Finally, data should be preserved in their original raw form. All data cleaning, summarization, gap-filling, analysis, and visualization should be done separate from the original data, preferably in a script so that all changes to the data are well documented and repeatable.

### 3.1.3 Assure

The goal of data assurance is to check that data are usable. The first steps in data checking should occur when the data are still fresh. In my lab field tally sheets are initialed by the person collecting the data, and that evening someone else independently checks (and initials) that all data sheets are accounted for, all values entered are clear, and data conform to expected values and codes.

As noted earlier, any quality assurance and quality control (QA/QC) should be documented and preferably recorded in reproducible computer scripts. Any values that are estimates, rather than raw observations, should be clearly identified. Data should be checked for adherence to predefined codes and acceptable minimum and maximum values. Many key variables will benefit from adding additional quality flag columns to indicate the level of reliability of records—this is preferable to deleting or "fixing" records. Files that have been through QA/QC should have this tagged in the file name. It is common for many agencies to release data at different data-processing levels, for example with level 0 being the raw observations, level 1 being the QA/QC'd data, and higher levels referring to more derived products (for example, gap-filled, spatial or temporally scaled, or processed through models).

### 3.1.4 Describe

To be usable beyond a short time after being collected, data need to be well documented. This data about the data, called *metadata*, is discussed in section 3.2. The goal of data description is ultimately to ensure that data can be used correctly by individuals who have had no direct contact with the data collectors.

### 3.1.5 Preserve

The goal of data preservation is to ensure that data are not lost over time. High-quality data often represent the life's work of dedicated ecologists—ensuring data are preserved and usable by future generations represents their scientific legacy irrespective of personal views on data sharing. However, data preservation cannot be put off to retirement! As almost everyone has learned the hard way, data should be backed up frequently so that information is not lost if files are corrupted or deleted, or if the machines they are on are lost, stolen, or destroyed. Beyond this, the long-term usefulness of data is greatly extended by depositing data in data repositories. Public, searchable repositories are obviously preferred, but as noted earlier some data has ethical or legal restrictions on data sharing (for example, human subjects, endangered species) and thus must be stored in long-term repositories that respect these restrictions. Data deposited in any repository should have a license specified, so any restrictions on reuse are clear. A multitude of possible data licenses already exist (for example, Creative Commons), and it is considerably easier on both data providers and users to work with standard, legally vetted licenses rather than making up ad hoc data policies.

### 3.1.6 Discover, Integrate, Analyze

While good data management ensures that we can integrate and analyze our own data, data deposited in public repositories with well-specified metadata should ultimately be discoverable and searchable by others. As noted earlier, analysis and forecasting are increasingly being done by researchers who are able to integrate many data sources. However, the landscape of available data is changing constantly, such that one real challenge is developing experience about what data repositories exist that are relevant to a specific question. Data integration is rarely automatic. Once discovered, data need to be checked, via DOIs or other identifiers, to ensure that data are consistent with metadata and quality checks and that the same data, found from two different sources, are not used twice. Finally, skills need to be developed on how to access and manipulate the data (section 3.3), integrate and process data in workflows (chapter 4), and analyze and visualize the results (which is the focus of the remainder of this book).

## 3.2 DATA STANDARDS AND METADATA

### 3.2.1 Metadata

Metadata is data about the data—it documents what the data are, how they are formatted and organized, as well as who, why, and how they were collected. As a general principle, metadata should contain enough information to make the data usable by someone who has never worked on the original project; it needs to document all the warts and wrinkles and caveats. Metadata is clearly critical for any data synthesis or forecasting, since it allows forecasters to determine whether the data contain what they are looking for and are appropriate for the analysis. It extends the lifetime of data, since use becomes far less reliant on the original data collectors, whose memory will always fade with time and will not always be accessible. As noted in the previous section, the first person we document data for is our future, forgetful self.

Metadata itself should include obvious things such as descriptions of all variable names (typically column headers), units, and any abbreviations or codes used in the data. As we will see in later chapters, it is also absolutely critical that estimates of uncertainties in the data be included, either on a value-by-value basis within the data itself or on a variable-by-variable basis in the metadata. Metadata also needs to be recorded for nontext data, such as images, videos, and other digital instrument data. For instrumental data this needs to include information on the times and methods of calibration and preferably also the raw calibration curves. While most users will not need this level of detail, it is generally safe to assume that new and better quantitative approaches will be developed in the future and that when it comes to synthesis, it is often necessary to reprocess data using standardized methods.

Other key components of metadata are where, when, and how data were collected. Geographic and temporal information is critical for higher-level synthesis, and understanding the methods used is often critical to interpreting data. Furthermore, methods need to be explained with the realization that what may be standard equipment and protocols today may not be standard in 10 to 20 years. Similarly, many labs will frequently say they use the same standard protocol, but each may have tweaked or tuned the approach slightly, such that no two are doing the exact same thing.

In addition to how the data were collected, it is important to also note why the data were collected and by whom. This should include who made the measurements, who entered them, and who to contact if there are questions. Metadata should also include the log of any changes to the data, including basic data cleaning and quality checks. Data scientists refer to this as the data's *provenance*.

### 3.2.2 Data Standards

One of the challenges in working with ecological data is the remarkably heterogeneous nature of the data. Ecologists measure a wide diversity of systems and processes using many different approaches, so some degree of heterogeneity is inevitable. On top of this, a large fraction of ecologists have historically been trained on projects that involved a relatively modest amount of data and a small number of collaborators. Because of this, the specific format that data was stored in for any given project was most often unique to that project. This is particularly true for data collected by hand and entered into spreadsheets, where the column headings can be very diverse among labs, and sometimes even among members of the same lab, even when collecting essentially the same data. In contrast to well-structured "big data," these unstructured, uncurated data are often referred to as the "long tail" because they represent the large number of small files in a rank-abundance distribution of file sizes (Palmer et al. 2007). To data managers these data are analogous to the abundance of rare species in a tropical forest. The challenge the long tail presents to synthesis is fairly obvious—it becomes very labor intensive to reconcile and merge data from different groups if every data set has a unique format. A variety of tools exists for trying to help automate this problem, but the challenge of inferring meaning from cryptic variable names, and ensuring that two groups really mean the same thing for a measurement is nontrivial.

Synthesis is greatly facilitated by the use of *data standards* where they exist. In the simplest form, standards provide a common definition of variable names, units, and abbreviations or other codes. They often involve working with a single common file format. Standards greatly increase the ease with which information can be searched and synthesized. For this reason it is also critical to adhere to *standards for metadata* in addition to standards for data itself. Storing metadata in a standard format (for example, Ecological Metadata Language), rather than an arbitrary README file or document, makes it considerably easier both for humans to find what they're looking for and for machines to parse out information. This can considerably speed up the processes of data discovery and integration.

The fixed vocabularies of standards (referred to as *ontologies*), for both data and metadata, also greatly reduce the ambiguity about the meaning of different variables and descriptions (respectively). Even if not working with data from others, *moving all data of a similar type to one standard greatly simplifies analysis, synthesis, and forecasting.* For some ecological data types there may be multiple competing standards for data or metadata, in which case interconversion is usually something that can be automated. For many types of ecological data there are no standards, in which case the act of producing a well-documented synthesis can often help establish a standard, especially if it engages multiple labs that discuss the ontology and can reach an agreement to use that standard moving forward. Outside of synthesis, at minimum adhering to metadata standards and SI units can greatly reduce the challenge of working with such data.

When standardizing data, a question often emerges whether to reduce all data to the lowest common denominator (for example, the intersection of what was measured by different groups) or to have a standard that includes everything that everyone could have measured (that is, the union). If the former, then information is inevitably lost, while the latter risks needing to revise the standard every time a new data set appears that has additional information not covered by the current standard (Jones et al. 2006). There is no correct answer to this, but as with most things the best course of action is usually some middle ground, such as defining a set of core variables, establishing a protocol for adding new variables, and adopting a file format that is flexible enough to accommodate ancillary data (see box 3.1)

## 3.3 HANDLING BIG DATA

Working with big data involves many advantages—big data usually have a well-documented data standard and come in a standard format. This standardization makes big data predictable and thus one can readily develop workflows for working with this data (chapter 4). The trade-off with big data is volume. This makes it harder to access big data, harder to process to find the part you're looking for, and harder to perform computations and analyses. The landscape of tools for dealing with big data is changing rapidly, so the following discussion will focus on the underlying concepts rather than specific tools.

What constitutes big data will be very different from field to field. Ecologists have long had the operational definition of "big data" as anything too big to open in Microsoft Excel. While somewhat tongue-in-cheek, this represents an important threshold for data users. Beyond this size, working with data requires learning new tools, very frequently including the need to learn some form of computer scripting language (for example, R, MATLAB, Python, Perl) to handle data processing, analysis, and visualization. I would argue that such skills are critical for modern ecologists, but they do have a steep learning curve. The move beyond spreadsheets means

### Box 3.1. Self-documenting Data

The bulk of ecological data are stored either in hand-entered spreadsheets or text files from data loggers. These data are easy to work with and reasonably portable, but generally require metadata and provenance to be stored in a separate file that's external to the data, and thus can potentially become separated from it. Databases solve many of these problems, as a database schema provides a lot of documentation about the structure of data and many database systems log any changes to the data. However, databases are less portable, as most systems require information to be dumped to external files and reimported somewhere else to move information. It is also often challenging to move data from one type of database to another. Alternatively, many task-specific pieces of software provide data files that are compact, portable, and contain metadata, but are specific to a particular piece of equipment.

Self-documenting files provide a middle ground. The key to these files is that the data and the metadata are stored together in the same file, but in a format that is flexible and extensible. They are also good at storing data of many shapes—such as individual constants, vectors of data, tables, multidimensional arrays, or more complex data structures (for example, geospatial polygons). In addition, self-documenting files can provide provenance by logging the history of file changes.

The two most common self-documenting file formats are the network common data format (netCDF), developed by the National Center for Atmospheric Research (NCAR), and the hierarchical data format (HDF), developed by the National Center for Supercomputing Applications (NCSA). In recent years the underlying file structure for these files has actually converged, such that tools to work with netCDF files will now work on HDF and vice versa. In addition to the self-documenting nature and flexibility of these file formats, they also benefit from being operating system independent and can be opened by a wide variety of software and programming languages. Because of this flexibility, these file formats are frequently used by large data providers for a number of their data products (for example, the National Ecological Observatory Network [NEON], NCAR, NASA, and Ameriflux).

that the user can't visually see all the data, which affects our mental model of the data set, such as how we see different variables as related to one another. This also changes how we perform QA/QC—for example, providing less opportunity to look deeply at individual outliers and relying more on graphs and summary statistics to identify invalid data and outliers. In other ways this change is for the better, as QA/QC is more likely to be documented in data processing scripts and data are less likely to be hand-manipulated in ways that are not easily repeatable.

While the Excel threshold is a very real one for ecological big data, it is not the only one. It is also gradually becoming the less important one, as the average data scripting skills among ecologists are increasing and the size of many data sets becomes even larger. The two other important "big data" thresholds are when data become too large to fit into a computer's active memory (RAM) all at once, and when they become too large to fit in the storage memory on an individual computer (which I'll refer to as *disk space*, as traditionally this was a hard drive). These are moving thresholds, as the capacity of RAM and disk space is ever increasing, but they never go away. The challenge of the RAM threshold is that computers are many orders of magnitude faster when performing computations on data that are in RAM than when they have to load data from storage. The RAM problem is occasionally one you can buy your way out of, simply by purchasing more RAM or increasingly by buying solid-state storage that is considerably faster than traditional hard drives. However, if this solution is not viable, then a slowdown is inevitable. Some RAM problems can be solved by dividing a big data set up into many smaller data sets that can be processed on many different computers (such as a computer cluster). Alternatively, there are special "external memory" computer libraries and algorithms that can considerably lessen the burden of running out of RAM by being very strategic about when data are read and written to hard drives (Vitter 2001; Arge et al. 2003). These algorithms can make the difference between an analysis slowing down versus grinding to a virtual standstill. That said, either of these approaches can involve a substantial investment of time and effort, as new tools need to be learned and your code will sometimes need to be completely rewritten.

Beyond the RAM threshold is the challenge of working with data that will not fit on a single computer. Ecologically, this problem is frequently encountered in remote sensing. For example the full LANDSAT data set requires over 1 petabyte of storage (1000 Tb). However, the spatial and temporal scope of such data is critical for many ecological forecasting problems. Working with data of this size requires a shift in how we think about computation. Traditionally, whether working on an individual laptop or a supercomputer, analysis has been done by bringing the data to the software that's being used for analysis and computation. The alternative paradigm, popularized by the MapReduce (Dean and Ghemawat 2008) paradigm and Hadoop software, involves moving computations to where the data are stored. In doing so subsets of the overall data set are distributed over a number of different computers, so that each subset of data fits in the storage (and preferably within the RAM) of each machine. Working in this paradigm requires that we are able to divide an overall analysis into parts, each performed on a data subset, and that we are able to combine the results from each subset into an overall conclusion. Not every problem fits within this paradigm, but fortunately many common analyses do. For example, most analyses of land cover can be done image by image. Similarly, most statistical analyses are fundamentally based on either a sum of squares or a sum of log likelihoods, and in either case we can perform that sum on each data subset independently (Dietze and Moorcroft 2011). It should be noted that this paradigm can also be an effective solution to RAM-limited problems as well. As with the large memory algorithms used for the RAM-limited problems, converting an existing problem to MapReduce-style approaches can require a substantial investment of time and rewriting if an analysis has not yet been adapted to these tools. Also, not all problems are as easily adapted to this paradigm. For example, analyses involving spatial or temporal covariances or other spatial interactions do not have this property of just being a simple sum, so they require more sophisticated approximations (Govindarajan et al. 2007).

In addition to the problem of working with big data, there are also a number of challenges with discovering, accessing, and downloading big data (Vitolo et al. 2015).

As with the tools for working with big data, the standards and protocols for accessing data through the Internet are likely to change radically over the lifetime of this book, but there are some overarching challenges. First is the tension between making data human accessible versus machine accessible. A web service that is highly user friendly when grabbing an individual data set here or there may perform horribly when that process needs to be automated to grab large volumes of data or update data frequently. Therefore, like with shifting from spreadsheets to scripts for looking at data, forecasters frequently need to shift from web browsers to scripts for accessing data. Thankfully, a number of simple command line tools, such as wget and curl, have stood the test of time. For example, in many cases downloading a file is as simple as typing *wget* and then the URL of the file, but more advanced options exist for dealing with authentication, downloading large sets of files, and so on. Increasingly scripting languages are also making pulling data off the web part of their default "under the hood" behavior. On the other side, there are web service architectures, such as REST and SOAP, that are much more powerful and machine-friendly, but incur a steeper learning curve. Unfortunately, the choice of what web services need to be learned is determined by the data provider, not the user! Thankfully, many of these services are such a ubiquitous part of Internet infrastructure that most have accessible tutorials.

## 3.4 KEY CONCEPTS

1. Adherence to data management best practices benefits ourselves and our collaborators, is often a legal requirement, and provides transparency to science. Data management should be a routine part of our daily workflow rather than an onerous post hoc headache.
2. Data management should be automated, should be documented, and should ensure that data are always recoverable.
3. Data life cycle: plan, collect, assure, describe, preserve, discover, integrate, and analyze.
4. Metadata ensures that data can be used correctly by individuals who have had no direct contact with the data collectors.
5. While ecological data are heterogeneous, moving all data of a similar type to one standard greatly simplifies analysis, synthesis, and forecasting.
6. The challenges of working with big data involve a number of key thresholds, each with its own learning curve: shifting from spreadsheets to scripting; running out of memory (RAM); and running out of storage (disk).

## 3.5 HANDS-ON ACTIVITIES

https://github.com/EcoForecast/EF_Activities/blob/master/Exercise_03_BigData.Rmd

- Pulling data directly off the web
- Web scraping
- grep, system, RegExp
- netCDF, wget
- SOAP
- cron

# 4

# Scientific Workflows and the Informatics of Model-Data Fusion

*Synopsis: Forecasts must be transparent and accountable, and generally they must be updated regularly as new data become available. In keeping with the theme of "best practices," we set the stage with the informatics of model-data fusion, before diving into the statistics, by describing tools and techniques that make science more transparent, repeatable, and automated. These tools are not isolated to forecasting—they should become part of all aspects of science—but they are an essential part of a forecaster's toolkit.*

Ecological research is being subject to increasing public scrutiny. Regardless of whether this scrutiny is politically motivated or in the form of completely legitimate requests to know the details behind policy-relevant science, it remains true that the bar has been raised on the need for transparency, accountability, and repeatability in science. As noted in the previous chapter on data, adhering to best practices not only makes our research more robust in the face of criticism, but it also leads to better science. The focus of this chapter will be on the best practices for the overall workflow for data synthesis, modeling, and forecasting.

## 4.1 TRANSPARENCY, ACCOUNTABILITY, AND REPEATABILITY

Reproducibility is one of the fundamental tenets of science. For any scientific result to be valid it should, at least in theory, be something that an independent team could repeat and find similar results. Furthermore, given the same exact data and tools, any independent team should be able to reproduce the same analyses. Unlike collecting data in the field or running an experiment in the lab, which will never produce the exact same data set two times, the results that come out of data synthesis, analysis, or forecasting should be easily and exactly repeatable. Transparent and repeatable science allows other researchers to verify results, to check assumptions, and to more easily build upon the research of others. Beyond philosophical ideals, this kind of reproducibility is particularly important pragmatically in forecasting, where analyses and projections need to be repeated automatically as new data become available.

As an example, take something as simple as repeating a statistical test. In practice, the specific software package used to generate a test statistic is often not reported in publications. If it is reported, that exact version of the software is often no longer available, and many systems are not backward compatible, meaning it's not clear that someone else repeating the same test on the same data would get the same exact answer (Ellison 2010). These concerns are increased when using proprietary or closed-source software, as the underlying operations are not verifiable. That said, this is definitely not a suggestion to code every analysis from scratch, as that leads to inefficient, redundant work that may be prone to bugs. Open software has the advantage of being verifiable, and, when employing popular tools, it can greatly increase the number of eyes that have looked for errors.

The challenges of repeatability increase rapidly as the complexity of an analysis increases. There is more room for error and inconsistency when either the number of steps or the complexity of each step increases. For example, computer modeling should be the most exactly replicable part of ecology, but in practice unless the specific version of the code and inputs used are archived, it is virtually impossible. With any complex analysis, model, or forecast, archiving the code and inputs used in a publicly accessible location (rather than "available upon request") is necessary to ensure openness and repeatability, just as much as archiving data was shown to be necessary in chapter 3. Research has shown that materials that are "available upon request" rapidly become inaccessible, vanishing at a rate of 17% per year (Vines et al. 2014). As with data management (chapter 3), both journals and funding agencies are increasingly requiring this kind of archiving. Thankfully, there are now a multitude of open software development tools and code repositories that make it possible to make not only specific versions of code available but also the overall development history (box 4.1).

## 4.2 WORKFLOWS AND AUTOMATION

To ensure replication is possible, in addition to archiving data and reporting the tools used, all of the steps of an analysis also have to be recorded. This *sequence of steps performed in an analysis* is referred to as the *workflow*.

Formally, a workflow should detail the exact sequence of steps that take a project from start to finish. The details of such workflows are commonly stored in the notebooks of the research team conducting an analysis. Even if well-organized and digital, these notes will only be human readable, which means that time and effort will be required to manually repeat an analysis and small details may be missing—things that are so obvious to the person conducting the analysis that they don't get written down, but which are a mystery to a new user. In addition, manual workflows like these are error prone. It's easy to forget steps, perform steps out of order, and encounter errors. When this happens, sections end up being repeated, leading to multiple versions of the same files, which can multiply confusion about what specific version of a file was used in an analysis. Generating numerous versions of files is particularly easy to do (figure 4.1), especially when developing or learning an analysis, and manual workflows do not provide the transparency and provenance tracking to ensure that analyses were actually done as intended. Even in my own lab, where we are very conscious of this challenge, I can assure you that we have hard drives full of manually numbered file versions and students have needed to repeat analyses because they used the wrong version of a file.

Increasingly, more formal scientific workflow tools and software are becoming available that track and record all the steps performed in an analysis (Ellison et al.

---

### Box 4.1. Software Development Tools

Software development is a multibillion-dollar industry. Furthermore, that industry is essential to most of the world's economies. However, when scientists start to write code they frequently try to reinvent the wheel on really basic issues like keeping track of versions of code, tracking bugs and proposed features, code testing, and project management. While it is unrealistic to ask ecologists to become experts in software development, a little time spent up front looking into the currently available tools and approaches can save enormous headaches later on. The following is an example software development pattern that highlights how different types of tools could be used, followed by a short glossary of terms:

1. At the start of the day, *pull* the latest code from the *mainline repository* to your local repository.
2. In an *issue tracking system*, look at what tasks are assigned to you or required for the next *milestone*. Open the specific issue or bug to be worked on. Within the issue tracker, discuss proposed changes with collaborators, solicit ideas, and explain design decisions.
3. Create a new *branch* of your local repository for this issue:
   a. At each incremental change, *commit* the code, referencing the name or number of the issue being worked on.
4. When the task is complete:
   a. Test that the code compiles and runs automated tests.
   b. *Push* the code back to the remote repository.
5. Submit a *pull request* that these changes be added to the mainline:
   a. Collaborators view the proposed code and discuss whether to accept, reject, or request additional changes.
   b. Once the code is accepted it is pulled into the *mainline*.
   c. All commit logs and discussions of the pull request and issue are permanently associated with those changes.
   d. The commit triggers a *continuous integration system*, which notifies all collaborators about success or failure of compilation and tests
5. Close the issue, delete the branch (both may occur automatically with pull request).
6. Once all issues for a milestone are complete, tag that revision and announce release to users.

BRANCH—A copy of the code, often created to work on a specific issue or development task.

COMMIT—To save a copy of the current branch into your local repository.

CONTINUOUS INTEGRATION SYSTEM—Software that checks that code in a repository compiles and passes a predefined set of tests. Such systems are often set to run whenever new code is added to the mainline.

ISSUE TRACKING SYSTEM—Software system that keeps track of reports of bugs, tasks, and requested features, which are collectively known as issues. Issue tracking is often integrated within version control.

MAINLINE—The main branch in the version control system. Also known as the trunk or master. Some developers will split the mainline into a development branch and a stable branch.

MILESTONE—A collection of grouped issues, usually leading to a specific deadline or the release of a new feature.

PULL—The process of merging code from one branch or repository into another.

PULL REQUEST—Proposal by a user to merge the user's code into another repository. Most often this is a request by a developer that a specific branch be pulled into the mainline. Pull requests have to be approved by the owner of the other repository.

PUSH—Sends changes in a local repository to a remote one. Often used to push changes in a local repository to the user's own remote repository in a cloud-based service. Occurs without the need for remote approval.

REPOSITORY—Collection of files, logs, and so on stored within the version control system, either locally or on a remote server. A repository will typically store multiple branches.

2006; Boose et al. 2007; Goble and Roure 2009). Many workflow tools, such as Kepler, present workflows in an intuitive graphical interface that captures the steps performed and the flows of information from step to step (Ludäscher et al. 2006). Many of these systems also track the full provenance of the workflow, creating a permanent record of the exact versions of data sets and tools used, outputs produced, and any logs and messages reported along the way.

Outside of formal workflow software, it is also possible to manually construct workflows that meet many of these same goals of automation, repeatability, and provenance. At the simplest level this could be a well-documented set of scripts that can repeat analyses and clear conventions for ensuring that steps are executed in the proper order (for example, script names that begin 00, 01, 02, and so on and/or a master script that calls all other steps). Revisions of such scripts should be tracked using version control software to ensure that a reproducible record exists of what versions of each step go together. A greater level of formality would be to combine such steps into an overall library or package within a specific language. Increasingly it is also possible to use a system that can combine executable code with text (R Markdown, Python notebooks, and so on) to ensure the reproducibility of whole documents, such as reports or manuscripts. Such systems make it far easier to ensure that documents contain the latest results. As noted in the previous section, it is important to capture not only the exact data and scripts used but also the specific version of software used. A variety of tools exist to do this, which range from software-specific tools for recording and retrieving versions (for example, the specific version of R and all R libraries used) up to capturing a snapshot of the whole operating system and all tools used (for example, virtual machines). While not always possible in practice, a good workflow should be executable, start to finish, by a third party on a different computer than the one where the analysis was initially conducted. This helps ensure

FIGURE 4.1. The importance of provenance and version control. Source http://xkcd.com
/1459/.

that all steps are properly documented and automated and that there are not "hidden" dependencies (for example, external files needed that are not documented, or
internal files that are needed but that are not being updated by the workflow). Continuous integration services exist that aim to automate such testing—for example, by
testing workflows on remote cloud servers.

Within forecasting, one of the key advantages of formal workflows is the automation of research tasks. Making a forecast "operational" is not merely a statistical
exercise in fusing models with data (chapters 13 and 14), but it's also an informatics
task of gathering and processing new data, managing the execution of analytical
workflows, and reporting results. The practical business of ecological forecasting
often results in a lion's share of effort being spent on informatics rather than statistics and modeling.

## 4.3 BEST PRACTICES FOR SCIENTIFIC COMPUTING

It should be clear already that ecological forecasting involves a combination of informatics, statistics, and modeling that requires some general familiarity with scientific
computing. This section summarizes some of the "best practices" for computing
(box 4.2), which have evolved from the overall experience of the much larger field of
software development. It also addresses some specific challenges that often face the
scientific community, such as the need or desire to make code public and open, and
a general lack of formal training in coding and software development.

It is often said that ecologists write horrible code. In my experience ecologists misinterpret this criticism and mistakenly think that the key to better coding is figuring
out how to do something faster or in fewer steps. It is not the elegance or efficiency
of code itself that is the biggest problem, but rather a more broad set of bad habits

## 4.2 Summary of Best Practices

The following recommendations are reproduced from Wilson et al. (2014):

1. Write programs for people, not computers.
   a. A program should not require its readers to hold more than a handful of facts in memory at once.
   b. Make names consistent, distinctive, and meaningful.
   c. Make code style and formatting consistent.
2. Let the computer do the work.
   a. Make the computer repeat tasks.
   b. Save recent commands in a file for reuse.
   c. Use a build tool to automate workflows.
3. Make incremental changes.
   a. Work in small steps with frequent feedback and course correction.
   b. Use a version control system.
   c. Put everything that has been created manually in version control.
4. Don't repeat yourself (or others).
   a. Every piece of data must have a single authoritative representation in the system.
   b. Modularize code rather than copying and pasting.
   c. Reuse code instead of rewriting it.
5. Plan for mistakes.
   a. Add assertions to programs to check their operation.
   b. Use an off-the-shelf unit testing library.
   c. Turn bugs into test cases.
   d. Use a symbolic debugger.
6. Optimize software only after it works correctly.
   a. Use a profiler to identify bottlenecks.
   b. Write code in the highest-level language possible.
7. Document design and purpose, not mechanics.
   a. Document interfaces and reasons, not implementations.
   b. Refactor code in preference to explaining how it works.
   c. Embed the documentation for a piece of software in that software.
8. Collaborate.
   a. Use premerge code reviews.
   b. Use pair programming when bringing someone new up to speed and when tackling particularly tricky problems.
   c. Use an issue tracking tool.

that surround our approach to scientific computing. Great code doesn't have to be fast, but it does have to be organized, clear, and reusable.

My top recommendation is that any code development, from simple data processing through sophisticated modeling and forecasting, should *start with a plan*. One of the real sources of "bad code" in ecology arises from starting to write code immediately and trying to solve problems as they come up. As with writing anything it is

```
## Define settings, load required libraries

## Read and organize data

## Define priors and initial conditions

## Enter MCMC loop

        ## propose new parameter value
        ## run model
        ## evaluate likelihood and prior
        ## accept or reject proposed value

## Statistical diagnostics

## Visualize outputs
```

FIGURE 4.2. Outline of a typical Bayesian statistical analysis.

helpful to start with a *flowchart* or *outline* of what you want the code to do. This should be a high-level description of the general steps involved, not a description of *how* each step will be implemented (figure 4.2).

At this high level, code should be designed to be *modular*. Individual actions should be encapsulated as functions, and the *inputs*, *outputs*, and *purpose* of each should be planned in advance and well documented before a single line of code is written. Once the conceptual purpose of a function is well defined, it is possible to change the internal implementation of any function completely without altering the overall behavior or the system. In addition to isolating tasks, one of the advantages of modularity is that it makes it easier to reuse code. In solving any coding problem there is a balance between meeting immediate short-term goals versus generalizing code so that a task is more efficient later. Learning this balance is difficult. For most ecologists the pattern is to err on the side of being too specific, which in the long run means that virtually the same code ends up being written time and time again, often by the same individual. For a more complex project, spending a few days discussing how the code should be structured can expose where challenges are likely to arise, allows code to be more easily generalizable, and results in far fewer ad hoc solutions to problems. From this high-level plan it is often useful to next sketch out *pseudo-code* that *describes how each step will be implemented*. Pseudo-code should be human readable but structured similarly to the code you intend to write (for example, loop over *x*, calculate *y*). Writing pseudo-code lets you work through high-level coding issues without having to sort out every detail, and will often serve both as the template for implementing the code as well as the first draft of documentation.

My second recommendation is to make the use of version control and other software development tools a strongly ingrained habit (see box 4.1). Version control provides a record of all changes to the code, along with comments describing the reason for the changes. Version control eliminates the profusion of hand-numbered file versions (see figure 4.1) and allows you to go back to previous versions. In particular, when working collaboratively version control and other development tools are all but essential. In general, each individual programming task should be saved to version control. Including multiple changes in each commit makes it much harder to track down individual bugs. This leads to my third recommendation, which is that changes should be introduced incrementally so that the code can be tested and errors identified quickly. Incremental changes also allow you to learn from each step and adjust your plan as needed.

Fourth, documentation is best located within the code. No one likes to spend time documenting code post hoc, but if you have begun with a plan as described earlier, the large majority of the documentation of the functions (inputs, outputs, purpose, behavior) and their implementation should be in place *before* you write a single line of code. If documentation exists in an external document, or in multiple places, it will always be out of date. A number of options exist (for example, Doxygen) for most programming languages to automatically extract documentation for comments within code, provided they follow certain conventions for tagging information. The tags in such systems are a good guide to the types of information that should be included in code documentation. In addition, when it comes to the nitty-gritty of code organization, a variety of style guides exist for most languages. More important than which style is used is that a group working together agree on a particular style. Furthermore, if code is well organized and uses meaningful variable names, the code itself should be fairly self-evident in terms of *what* it does, and *documentation can instead focus on the purpose and design*. For example, a comment such as "loop over parameters" typically restates the obvious, but what won't be obvious to someone else is which parameters are being looped over, why you're looping over them, and what's being calculated in the loop. As with metadata on data, never forget that code documentation is first and foremost written for your future self, who will remember far less about what a specific piece of code is supposed to do, and why certain design choices were made, than your present self assumes.

My final recommendation is to never forget that the primary goal is to get science done, not to produce beautiful code. To repeat Sheryl Sandberg's paraphrase of Voltaire, "Done is better than perfect." The first goal is to have something that works. Only once that is in hand does it make sense to focus on issues of refining and optimizing code.

## 4.4 KEY CONCEPTS

1. The results that come out of data synthesis, analysis, or forecasting should be easily and exactly repeatable.
2. Workflows define the exact sequence of steps performed in an analysis.
3. The practical business of ecological forecasting often results in a lion's share of effort being spent on informatics rather than statistics and modeling.
4. Any code development should start with a plan.
5. Code should be modular, with the inputs, outputs, and purpose of each part designed in advance. Modularity isolates tasks and enables reuse.
6. Make the use of version control and other software development tools a strongly ingrained habit.
7. Changes should be introduced incrementally so that the code can be tested and errors identified quickly.
8. Documentation is best located within the code and should focus on purpose and design.

## 4.5 HANDS-ON ACTIVITIES

https://github.com/EcoForecast/EF_Activities/blob/master/Exercise_04_PairCoding
.Rmd

- Version control
- Pair coding
- Code planning and modularity

# 5

## Introduction to Bayes

*Synopsis: This chapter provides a primer to readers who may not have had much prior exposure to Bayesian concepts and methods. For those with previous experience it might serve as a refresher or could be skipped. By contrast, those with no prior exposure to Bayes should focus on the concepts presented here, but will likely need to consult other resources to be able to perform analyses.*

### 5.1 CONFRONTING MODELS WITH DATA

In their influential book, *The Ecological Detective*, Hilborn and Mangle introduced much of the ecological community to Maximum Likelihood and Bayesian statistics (Hilborn and Mangel 1997). The subtitle for that book, "Confronting Models with Data," is an apt description of the central tenet of most statistical modeling: that data is the ultimate arbiter when fitting models and judging between alternative models. Statistical modeling assumes that you have one or more models and the goal is to fit these models to data (also known as *calibration or inverse modeling*), with any hypothesis testing occurring as a choice between alternative models. As we shift focus from ecoinformatics (chapters 3 and 4) to statistics, this "calibration" perspective will be the focus for the next few chapters. For purely statistical models this perspective will be fairly familiar—for example, you may believe the relationship between $X$ and $Y$ is approximately linear, but want to know the slope and intercept. This same perspective applies to process-based models, where most parameters in ecological models are empirically estimated rather than being well-defined physical constants. However, for process models we may have more prior information about the likely values of parameters, either because they are directly measurable quantities (for example, stoichiometric ratios) or because of prior experience with the model.

This chapter provides a quick overview of Maximum Likelihood and Bayesian methods, shows how we can use a Bayesian approach to incorporate this prior information about model parameters, and discusses why we would want to do so. It also introduces the important idea of using probability distributions to represent uncertainties in ecological forecasts. However, this one chapter cannot expect to substitute for an entire course or textbook on Bayesian statistics. In this way ecological forecasting is analogous to multivariate, time-series, or spatial statistics—it is a special topic that builds upon a general statistics foundation rather than being a substitute. Many such general books exist in Bayesian statistics, and a number of these are summarized at the end of this chapter. That said, forecasting does rely on Bayesian concepts, and thus this chapter lays the foundation for later chapters that discuss how to characterize and partition sources of uncertainty (chapter 6); fuse multiple data sources (chapter 9); and propagate, analyze, and reduce uncertainties (chapter 11). While we will retain the Bayesian perspective throughout the book, ultimately, in chapters 13 and 14, we will move beyond the paradigm of confronting models with data, and instead will think of ecological forecasting as "fusing models and data." However, this model-fitting perspective is useful to keep in mind throughout the following discussion.

## 5.2 PROBABILITY 101

An overarching premise of this book is that *the best way to combine models and data is through the use of probabilities*. Probabilities are used throughout this book—for example, to assess the likelihoods of data, the probabilities of model parameters and inputs, the probabilities of different ecological outcomes, and the risk associated with different management and policy actions. *The goal of this section is to review a few basic concepts in probability and the notation used throughout the book.*

The first key concept is that of a *random variable*, X. Instead of having a single fixed value, a random variable can take on multiple values, $x_i$, each with probability $p_i$. Symbolically, we write this as $P(X = x_i) = p_i$, and read this as "the random variable X takes on the value $x_i$ with probability $p_i$." While it is possible to specify these probabilities individually, it is more common to specify them using functions known as *probability distributions*. There are two types of probability distributions, discrete and continuous, depending on whether the random variable takes on only discrete values (for example, integers, categorical variables) or whether it is continuous. An example of a discrete random variable might be the roll of a dice, while a continuous random variable might describe the heights of students in a class.

For discrete probability distributions the *probability mass function (PMF)*, $f(x_i)$, specifies the probabilities, $p_i$, associated with each $x_i$. By definition, each individual probability must be nonnegative and no greater than 1, and the sum of all probabilities must equal 1. For example, for a standard 6-sided dice, the probabilities for $x = \{1,2,3,4,5,6\}$ are all $p_i = 1/6$, while $p_i = 0$ of $x < 1$ and $x > 6$. From each probability mass function we can also define a *cumulative distribution function (CDF)*, which is the probability that the random variable X is less than or equal to $x_i$, $F(x_i) = P(X \leq x_i) = \sum f(x)$. For a dice the CDF would be {1/6, 2/6, 3/6, 4/6, 5/6, 1}. Because probabilities must be nonnegative and sum to 1, the CDF will always start at zero and increase to a maximum of 1. Common examples of discrete probability distributions are the binomial, which describes the expected number of successes from $n$ independent trials (for example, coin flips) each with probability $p$, and the Poisson, which describes the expected number of events in time or space given some mean rate $\lambda$ (for example, number of individuals in a plot, eggs in a clutch, or disturbances in a year).

For continuous distributions we can similarly define a cumulative distribution function, $F(x_i) = P(X \leq x_i)$, and as with the discrete case, all CDFs start at zero probability and increase up to a maximum of 1, since the total probability cannot be greater than 1 and all probabilities have to be nonnegative. Analogous to the PMF, continuous distributions have a *probability density function (PDF)*, which is defined as the derivative of the CDF, $f(x) = dF/dx$. Therefore, just as the discrete CDF was the sum of the PMF, the continuous CDF is the integral of the PDF. As with the PMF, the PDF must be nonnegative; however, unlike the PMF, the PDF can be greater than 1 provided that the total integral of the PDF is equal to 1. For example, a uniform PDF between 0 and ½ will have a density of 2 on that range, since the integral has to equal 1 (which in this case is just the area of the rectangle, ½ × 2). This uniform PDF would then have a density of 0 at all values < 0 or > 1/2. Its CDF would be 0 for all $x < 0$, 1 for all $x > ½$, and in between would be a straight diagonal line from 0,0 to ½,1. When $x$ is a continuous number the probability of any specific $x_i$ is infinitesimally small, so probabilities are defined based on the area under the curve over some region of interest, $P(x_i \leq X \leq x_i + dx) = \int_{x_i}^{x_i+dx} f(x) = F(x_i + dx) - F(x_i)$. Examples of common PDFs are the Gaussian distribution (also known as the Normal) and the uniform distribution.

The other key concept in probability is how we describe the interactions of multiple random variables. To begin, we can define the *joint probability* of more than one random variable, $P(x,y)$, which is the probability of both $x$ and $y$ occurring simultaneously. It should be noted that the $P(x,y)$ is in general not equal to the product of $P(x)$ and $P(y)$ except for the special case when $x$ and $y$ are *independent*. Based on the joint distribution we can define two other probabilities that we will use throughout this book. The first and most important is the *conditional probability*, $P(x|y)$ (typically read as *the probability of* x *given* y), which is the probability distribution of the random variable $x$ conditioned on $y$ taking on a specific (fixed) value. The second is the *marginal probability*, $P(x)$, which is the probability of $x$ having integrated the joint probability over all possible states of $y$,

$$P(x) = \int P(x,y)dy \tag{5.1}$$

Finally, it can be shown that the joint, conditional, and marginal probabilities are related to each other according to the relationship

$$P(x,y) = P(x|y)P(y) \tag{5.2}$$

which says that the joint probability of $x$ and $y$ is equal to the conditional probability of $x$ given $y$, times the marginal probability of $y$.

The relationships between these different probabilities can be illustrated by a simple example of discrete probabilities, though bear in mind that these rules hold for all PDFs and PMFs. Consider the simple set illustrated by the Venn diagram in figure 5.1, where there are 10 points distributed across the set of $X$ and $Y$. The marginal

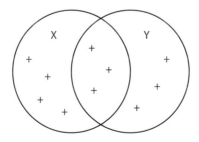

FIGURE 5.1. Graphical illustration of joint, conditional, and marginal probability.

probability of being in just $X$ is $P(X) = 7/10$, while the marginal of being just in $Y$ is $P(Y) = 6/10$. The joint probability of being in both $X$ and $Y$ is described by the intersection of $X$ and $Y$, $P(X,Y) = 3/10$. If we consider just the points in the set $Y$, the conditional probability of being in $X$ given that we are in $Y$ is $P(X|Y) = 3/6$. On the other hand, the conditional probability of being in $Y$ given $X$ is $P(Y|X) = 3/7$. We can also verify equation 5.2, since $P(X,Y) = P(X|Y)P(Y) = 3/6 \cdot 6/10 = 3/10$. Similarly $P(X,Y) = P(Y|X)P(X) = 3/7 \cdot 7/10 = 3/10$.

## 5.3 THE LIKELIHOOD

So how do we use these basic principles of probability to bring models and data together? Consider the simple case in figure 5.2, which depicts an observed data point at $x = 1.5$, and two alternative hypotheses, H1 and H2, which state that the true mean is either H1 = 0 (with a standard deviation of 0.7) or H2 = 3 (with a standard deviation of 1.3). Given this one data point, we can ask which hypothesis is more likely to have generated the observed data. Intuitively, H2 would be more likely because the probability density at $x = 1.5$ is higher for the right-hand curve (0.16) than the left-hand curve (0.06). More formally, we could calculate the actual probability within some small $dx$ around $x = 1.5$, but if we are just comparing the two hypotheses this $dx$ would cancel out and we still conclude that it is 2.7× more likely (0.16/0.06) that the observed data came from H2 than H1.

The intuition illustrated in figure 5.2 is formally expressed by the *likelihood principle*, which states that *a parameter value is more likely than another if it is the one for which the data are more probable*. In this simple case we considered only two alternative parameter values, H1 and H2, but in most cases we are considering a whole continuum of alternative parameter values. For example, one set of hypotheses might be that the mean, $\mu$, could be any value from −1 to 4; another might consider any real number. Given this set of parameters, a classic statistical goal is to determine the most likely parameter value, which is done using an approach called *maximum likelihood estimation (MLE)*: the most likely parameter value is the one that maximizes the likelihood. It should be noted that MLE is not Bayesian; it is a standard tool from classical (also known as frequentist) statistics, and all classic parametric statistics (ANOVA, regression, and so on) can be derived from this perspective.

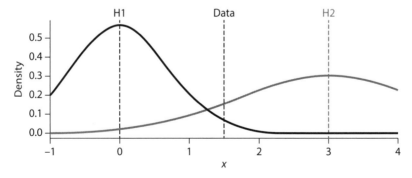

FIGURE 5.2. Probability of observed data point being generated under two alternative hypotheses.

For simple problems maximizing the likelihood can be done using the standard calculus-based approach that we would use to maximize any function:

1. Write down the likelihood function.
2. Take the derivative with respect to each parameter in the model.
3. Set the equation(s) equal to zero.
4. Solve for the parameter value(s) that maximize the likelihood.

In the case of our simple example we might assume that the data follow a Normal distribution ($N$), in which case the likelihood becomes

$$L = P(x = 1.5|\mu,\sigma^2) = N(1.5|\mu,\sigma^2) \tag{5.3}$$

It is worth noting that the likelihood, $L$, is expressed as a conditional probability. This conditional probability expresses the probability of the data conditioned on specific values of the model parameters, and thus the likelihood is commonly referred to as the probability of the data given the model, P(data|model).

For equation 5.3 we can write out the equation for the Normal, which gives

$$L = N(x|\mu,\sigma^2) = \frac{1}{\sqrt{2\pi\sigma^2}} e^{-\frac{(x-\mu)^2}{2\sigma^2}}$$

Because it generally makes the problems easier to solve, both analytically and numerically, but does not change the location of the MLE, it is common practice to take the natural log of the likelihood function before taking the derivative. It is also typical to multiply the log likelihood by $-1$ so that we are minimizing the negative log likelihood rather than maximizing the likelihood.

$$-\ln L = \frac{1}{2}\ln 2\pi + \ln\sigma + \frac{(x-\mu)^2}{2\sigma^2}$$

Let's next solve for the MLE of $\mu$ by taking the derivative and setting it equal to 0.

$$-\frac{\partial \ln L}{\partial \mu} = -\frac{(x-\mu)}{\sigma^2} = 0$$

$$\mu_{MLE} = x$$

Graphically this can be shown by plotting the $-\ln L$ against $\mu$ (figure 5.3), from which it is clear that if we observe only one data point, the most likely value of the mean is the same as that one observation. It is straightforward to extend this derivation to multiple independent data points, in which the likelihood becomes a product of Normal distributions

$$L = \prod_{i=1}^{n} N(x_i|\mu,\sigma^2) = \prod_{i=1}^{n}\frac{1}{\sqrt{2\pi\sigma^2}} e^{-\frac{(x_i-\mu)^2}{2\sigma^2}}$$

Applying the same steps as earlier, we first find the $-\ln L$, which has the advantage of turning the product into a sum,

$$-\ln L = \frac{n}{2}\ln 2\pi + n\ln\sigma + \sum_{i=1}^{n}\frac{(x_i-\mu)^2}{2\sigma^2}$$

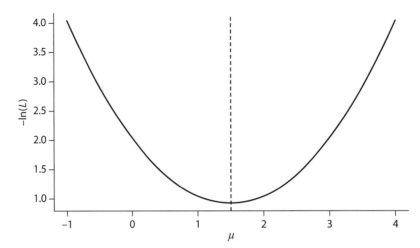

FIGURE 5.3. Negative log-likelihood profile for a Normal mean, $\mu$ (equation 5.3), assuming a known variance, $\sigma^2 = 1$, and one observation, $x = 1.5$ (vertical dashed line).

which is much easier to take the derivative of

$$-\frac{\partial \ln L}{\partial \mu} = -\sum_{i=1}^{n} \frac{x_i - \mu}{\sigma^2} = 0$$

Solving this for $\mu$

$$\sum_{i=1}^{n} x_i = n \cdot \mu$$

$$\mu = \frac{\sum_{i=1}^{n} x_i}{n}$$

shows that the MLE for $\mu$ is simply the arithmetic mean of the data. The same analytical approach can be applied to more complex models, such as a straight line with Normal error, from which it is easy to derive the standard equations used in linear regression. I have said before (chapter 2) that a model can be anything from a simple mean to a complex simulation model. In keeping with that it is worth emphasizing that the concept of likelihood can be applied equally to all models regardless of complexity or whether they are "statistical" models or "mechanistic" models. That said, for more complex models it is often impractical or impossible to solve for a MLE analytically, in which case it is standard practice to rely on numerical optimization algorithms to find the minimum of the $-\ln L$.

For most ecologists first encountering MLE, the concept of maximum likelihood is relatively straightforward, but they are intimidated by solving for the MLE, either analytically or numerically. However, in practice, what turns out to be most challenging isn't the calculus or code (which quickly becomes formulaic with a bit of practice) but knowing what likelihood to write down. To begin to address this let's divide the overall likelihood into two parts, which we'll call the process model and the data model. The *process model* is the part we're most accustomed to thinking about as "modeling"—it's the part of the likelihood that describes the model prediction for any particular set of inputs or covariates. Frequently, the process model is a determin-

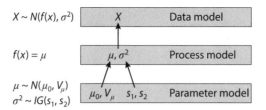

$X \sim N(f(x), \sigma^2)$

$f(x) = \mu$

$\mu \sim N(\mu_0, V_\mu)$
$\sigma^2 \sim IG(s_1, s_2)$

FIGURE 5.4. Graphical representation of data, process, and parameter models for a simple Normal mean. The process model describes the expected mean, which in this case is just a constant value. The data model describes the variability in the data around the process model. The parameter model (also known as prior) applies only to Bayesian models and describes uncertainty about the model parameters.

istic equation or simulation that describes the expected mean around which data are distributed. In our simple example earlier the process model is just the mean, $f(x) = \mu$ (figure 5.4), while for a linear regression, the process model would be the equation for the line, $f(x) = \beta_0 + \beta_1 x$. More complex process models might involve multiple simultaneous nonlinear functions or computer simulations. The *data model*, on the other hand, is the probabilistic portion of the likelihood that describes what we typically think of as the residuals—the mismatch between the model and the data. However, as we will end up using probability to partition many types of uncertainty and variability (chapter 2), the data model will not be the *only* probabilistic portion of the likelihood. More properly, the data model is best thought of as an attempt to capture the complexities and uncertainties in the data observation process. As we will see (chapters 6 and 9), data models can be used to capture multiple layers of complexity associated with sampling variability, random and systematic measurement errors, instrument calibration, proxy data, unequal variances, and missing data. That said, it is best to start simple when constructing a data model and to add such complexity incrementally. For example, in both the simple mean (earlier) and in a classic regression the data model is just the Normal distribution describing the residuals.

Before beginning to construct data models it is worth spending some time getting to know common probability distributions and what they are typically used for. Sometimes probability distributions are chosen because they "fit well," while at other times they are chosen because they are a natural description of the process (for example, Binomial coin flips, Poisson sampling variability, Exponential waiting times, Beta distributed proportions). A selection of common distributions are discussed in more detail in section 6.1.

The following are some of the questions that I often ask when specifying the data model:

First, is the data collection process continuous or discrete? In general it is best to use discrete PMFs for discrete data (for example, counts). It should be noted that a number of data types that are written as continuous numbers (for example, density) are often based on dividing discrete count data by a fixed area or time, and thus should be treated as discrete. Also worth noting, for discrete data, is whether the data are categorical, Boolean (true/false), counts, or some other integer data.

Second, are there range restrictions on the data? (For example, do the data have to be nonnegative? Is there a well-defined upper bound?) In general one should choose probability distributions that respect such boundaries, especially if the observed data are near the boundary. For example, variances should always be modeled with non-negative distributions, since squared errors can never be negative. That said, it is often fine to choose an unbounded PDF for continuous data provided the CDF outside the bounds is very low (for example, using a Normal PDF when all data are far from the boundary). Similarly, when the sample size for bounded, discrete data is large, it is not uncommon to approximate the PMF with a continuous PDF provided the data are not near any boundaries (for example, count data are nonnegative) and the PDF provides a good match. A common example is treating Binomial or Poisson data as Normal when the numbers involved are large.

Third, what process(es) generated the data? Understanding where data come from, and how they were collected, can be invaluable for accurately representing the complexities of real-world data. If you are using data you didn't collect yourself, it is advisable to spend some time talking through how you plan to model the data with those who collected it. If this is not possible, then it is very important to read all the details of the metadata (chapter 3), to read up on the methods used, and to talk to others who have collected similar data. Consider, for example, data on tree diameter growth. Intuitively, we know that diameter growth must be nonnegative, but if growth is inferred from repeated measurements of diameter breast height (DBH) with a DBH tape, then it's not unusual for the difference between those diameters to show shrinkage. This is because the actual data are for diameter, not growth, and diameter data have a high observation error. If, for example, we assume observation error to be Gaussian and independent in time, then even if the true process is nonnegative, the difference between these two random variables may often be negative. In this instance removing the negative values would lead to a systematically biased growth estimate. A data model that explicitly accounts for DBH measurement error would be preferable. If, by contrast, growth was measured using increment cores, then negative growth will never be observed. However, it is possible for tree cores to have missing rings, so the data model might want to account for the probability of having a missing ring. The overall growth model might then be split into a model of the probability of no growth and a model of wood growth conditional on not having a missing ring, each of which might be functions of different environmental variables. Finally, growth measured by dendrometer bands has low measurement error, and like tree rings is a direct measure of growth increment, but negative measurements also occur due to plant hydraulics. Dendrometer band data models must also account for systematic errors associated with the thermal expansion of the band, and possibly biases due to the pressure of the band itself. As this example illustrates, what appears to be one type of measurement, diameter growth, can lead to very different data models depending on the details of the data generation process.

Finally, what are the dominant sources of variability and uncertainty in the data-model mismatch? Common sources include sampling variability, measurement error, and process variability. Process variability may be random or, as we will discuss in chapter 6, it may show patterns in space or time, may vary systematically at the individual, plot, or site scale, or may be phylogenetically structured. Overall, the construction of a good data model involves much more than assuming independent Normal distributions, and one should work hard to avoid this "Gaussian reflex." Constructing data models requires thinking deeply about both your data and your model and digging into the details of the data generating process. This effort is rewarded with considerably greater statistical flexibility and ecological realism.

## 5.4 BAYES' THEOREM

The previous section illustrated the maximum likelihood approach, which was fundamentally based on estimating the probability of the data, $X$, given the model parameter, $\theta$, $P(X|\theta)$. While MLE is based on probability theory, the likelihood profile (figure 5.3) is not itself a probability distribution because it is generated by varying the "fixed" parameter that you are conditioning on, $\theta$, not the random variable, $X$. Because it isn't a PDF, the likelihood profile doesn't integrate to 1 (and no, you can't just normalize it to make it a PDF). This means that we cannot interpret the likelihood profile in terms of its mean, variance, quantiles, or any other statistics about its central tendency or uncertainty. More generally, given our aim to represent uncertainties based on probabilities, what we are really interested in finding is the probability distribution of the parameters conditioned on the data, $P(\theta|X)$, not the probability of the data given the model, $P(X|\theta)$.

If we return to our probability rules from section 5.2, it is possible to derive $P(\theta|X)$ as a function of our likelihood, $P(X|\theta)$. Specifically, from equation 5.2 we can write down

$$P(X,\theta) = P(X|\theta)P(\theta)$$

and

$$P(\theta,X) = P(\theta|X)P(X)$$

Because these joint probabilities are the same, $P(X,\theta) = P(\theta,X)$, we can combine these as

$$P(\theta|X)P(X) = P(X|\theta)P(\theta)$$

Solving for $P(\theta|X)$ then gives Bayes' theorem:

$$\underbrace{P(\theta|X)}_{posterior} = \frac{\overbrace{P(X|\theta)}^{likelihood}\ \overbrace{P(\theta)}^{prior}}{P(X)} \tag{5.4}$$

This simple derivation shows that the conditional probability of the parameters, which we refer to as the *posterior*, is related to the likelihood of the data by the *prior* probability of the parameters, $P(\theta)$. Based on our definition of marginal probability (equation 5.1), we can express the term in the denominator as

$$P(\theta|X) = \frac{P(X|\theta)P(\theta)}{\int P(X|\theta)P(\theta)} \tag{5.5}$$

from which it becomes clear that this term serves as the normalizing constant, ensuring that the posterior probability is a proper PDF that integrates to 1.

So how can we use Bayes' theorem to perform inference? Let's begin with a simple example involving individual probabilities rather than PDFs. In the following comic (figure 5.5) it is proposed that a neutrino detector measures whether the sun has gone nova but then lies to us with probability 1/36. The observed data, which is the answer the detector returns, is $X = YES$, and there are two alternative hypotheses for $\theta$, $\theta_1$ = nova and $\theta_2$ = no nova. From this we can derive two likelihoods:

$$P(X = YES|\theta_1 = nova) = 35/36$$

$$P(X = YES|\theta_2 = not\ nova) = 1/36$$

FIGURE 5.5. "Frequentists versus Bayesians." Source: https://xkcd.com/1132/.

As the comic notes, at the classic frequentist $\alpha = 0.05$ level, $0.05 > 1/36$, and thus this constitutes support for $\theta_1$, the sun has gone nova. However, what we're interested in is actually the posterior, $P(nova|YES)$, the probability that the sun has gone nova given that the detector says *YES*. To construct this probability we need to be able to specify the prior, $P(\theta = nova)$, which is our belief that the sun has exploded *prior to observing the data*. There are a wide range of ways that the prior could be constructed, varying from personal experience to astronomical theory, but in any case it's safe to assume that that this probability is pretty low. For the sake of discussion let's assume a value of $P(\theta_1) = 1/10{,}000$. If we plug these numbers into Bayes' theorem, we get

$$P(\theta_1|YES) = \frac{P(YES|\theta_1)P(\theta_1)}{P(YES|\theta_1)P(\theta_1) + P(YES|\theta_2)P(\theta_2)}$$

and substituting in our probabilities gives

$$P(\theta_1|YES) = \frac{\dfrac{35}{36}\dfrac{1}{10000}}{\dfrac{35}{36}\dfrac{1}{10000} + \dfrac{1}{36}\dfrac{9999}{10000}} = \frac{35}{10034} \approx 0.35\%$$

In this case we are a good bit more skeptical about whether the detector is lying to us or not. That said, in comparing our posterior to our prior we are now 35× more likely to believe that the sun has exploded than we were before observing the data. In other words, the data has had an impact on our beliefs, it's just that this evidence is not sufficient to overturn our prior experience. As Carl Sagan famously said, "extraordinary claims require extraordinary evidence." More generally, Bayes' theorem just formalizes in probability theory how most science is done: new evidence is never considered in isolation but must be viewed in the context of previous data. Indeed, one incredibly useful feature of Bayes' theorem is that it provides an inherently iterative approach to inference—*the posterior from one analysis can quite naturally become the prior for the next.*

Coming back to our more general problem of confronting models with data, let's next look at how we can use Bayes' theorem to estimate model parameters. Consider our earlier example where we were looking at the Gaussian likelihood that $X = 1.5$ given some mean parameter $\mu$ (equation 5.3). For the moment let's assume that $\sigma^2$, the observation error, is known. Let's also assume that our prior estimate of $\mu$, before we observed $X$, could also be described by a Gaussian distribution with mean $\mu_0$ and variance $\tau$, $P(\mu) = N(\mu|\mu_0,\tau)$. Because the denominator in Bayes' theorem is just a normalizing constant (and thus doesn't vary with $\mu$), our posterior is then proportional to (written as $\propto$) the likelihood times the prior

$$P(\mu|X) \propto P(X|\mu)P(\mu) = N(X|\mu,\sigma^2)N(\mu|\mu_0,\tau) \tag{5.6}$$

Solving this (see box 5.1) results in a posterior that is also a Normal distribution with mean and variance given by

$$P(\mu|X) = N\left(\mu \left| \frac{\dfrac{X}{\sigma^2} + \dfrac{\mu_0}{\tau}}{\dfrac{1}{\sigma^2} + \dfrac{1}{\tau}}, \frac{1}{\dfrac{1}{\sigma^2} + \dfrac{1}{\tau}} \right.\right)$$

This equation looks a bit unwieldy, but can be made much more manageable if we express our Normal distribution in terms of precisions (= 1/variance) rather than variances. If we define the prior precision as $T = 1/\tau$ and the data precision as $S = 1/\sigma^2$, then the preceding simplifies to

$$P(\mu|X) = N\left(\frac{S}{S+T}X + \frac{T}{S+T}\mu_0, S+T\right) \tag{5.7}$$

What this shows is that the posterior precision is simply the sum of the prior precision and the data precision, $S + T$. It also shows that the posterior mean is a weighted average between the prior mean, $\mu_0$, and the data, $X$, where the weights are determined by the precisions.

While this exact analytical solution applies only to the specific case of a Normal posterior arising from a Normal likelihood times a Normal prior, qualitatively the behavior it displays is common across most Bayesian analyses. Specifically, our *posterior mean is going to represent a combination of our new data and our prior information, where the impact of each is determined by their relative uncertainties.* If we are very confident about our prior inference, then the prior will have a lot of weight in the posterior. If we have lower confidence about our prior, then the data will tend

## Box 5.1. Normal-Normal Conjugacy

In equation 5.6 we saw a posterior that was proportional to a Normal likelihood times Normal prior

$$P(\mu|X) \propto P(X|\mu)P(\mu) = N(X|\mu, \sigma^2)N(\mu|\mu_0, \tau)$$

If we write out the Gaussian PDFs and drop all normalizing constants this becomes

$$P(\mu|X) \propto exp\left(-\frac{(X-\mu)^2}{\sigma^2}\right)exp\left(-\frac{(\mu-\mu_0)^2}{\tau}\right)$$

Expanding out each quadratic gives

$$= exp\left(-\frac{X^2 - 2X\mu + \mu^2}{\sigma^2}\right)exp\left(-\frac{\mu^2 - 2\mu\mu_0 + \mu_0^2}{\tau}\right)$$

$$= exp\left(-\frac{X^2 - 2X\mu + \mu^2}{\sigma^2} - \frac{\mu^2 - 2\mu\mu_0 + \mu_0^2}{\tau}\right)$$

Since we want to write a posterior probability for $\mu|X$, we then collect terms with respect to $\mu$

$$= exp\left(-\mu^2\left[\frac{1}{\sigma^2} + \frac{1}{\tau}\right] + 2\mu\left[\frac{X}{\sigma^2} + \frac{\mu_0}{\tau}\right] - \left[\frac{X^2}{\sigma^2} + \frac{\mu_0^2}{\tau}\right]\right)$$

From here we could complete the square and then solve for the normalizing constant. However, in this case we can more easily note that the preceding equation is going to have the same form as each of the two Gaussians expanded earlier and thus we can recognize that the posterior distribution is going to follow a Gaussian distribution as well. We can then find the mean and the variance by matching terms, recognizing that $\mu$ is the random variable. In the likelihood the random variable was $X$ and we can see that the $-X^2$ term is divided by variance, so matching terms, the posterior variance will be

$$\frac{1}{var} = \left[\frac{1}{\sigma^2} + \frac{1}{\tau}\right]$$

Similarly the $-2X$ term is multiplied by the mean/variance, and therefore the posterior will be

$$\frac{mean}{var} = \left[\frac{X}{\sigma^2} + \frac{\mu^0}{\tau}\right]$$

Solving these and plugging them into the Normal gives

$$P(\mu|X) = N\left(\frac{\frac{X}{\sigma^2} + \frac{\mu_0}{\tau}}{\frac{1}{\sigma^2} + \frac{1}{\tau}}, \frac{1}{\frac{1}{\sigma^2} + \frac{1}{\tau}}\right)$$

This derivation is an example of a more common phenomena known as *conjugacy*, which is when a posterior distribution is of the same functional form (that is, the same named probability distribution) as the prior. There are a wide range of conjugate likelihood-prior combinations that are in common use in Bayesian statistics, which can be found on the web or in any Bayesian textbook. For example, if we assume the same Normal likelihood on our data, $N(X|\mu, \sigma^2)$, and combine this with a Gamma prior on the precision

$$\frac{1}{\sigma^2} \sim Gamma(a, r)$$

then the posterior distribution of the precision also follows a Gamma distribution

$$P\left(\frac{1}{\sigma^2}\Big|X\right) = Gamma\left(a + \frac{n}{2}, r + \frac{1}{2}\sum(x_i - \mu)^2\right) \tag{5.8}$$

where $n$ is the sample size and the summation is just the sum of squares error.

Conjugacy confers two major advantages in Bayesian statistics. First, they correspond to cases that have analytical solutions that are well known and easy to look up. As we'll see in section 5.6, most Bayesian models don't have analytical solutions and need to be solved numerically. The second advantage is that we can sample directly from conjugate distributions (that is, draw random numbers from them). Therefore, even in complex Bayesian models with many parts, if the *conditional* posterior distribution of a parameter (conditioned on holding all other parameters fixed) is conjugate, then we can sample from that parameter directly—an approach known as *Gibbs sampling*. If all else is equal, choosing priors that are conjugate with your likelihood will increase the efficiency and convergence of Bayesian numerical methods.

to dominate through the likelihood. Similarly, given enough data, the likelihood will generally overcome any prior beliefs and dominate the posterior inference. The other general pattern in equation 5.7 concerns the posterior uncertainty: *the posterior distribution tends to be more precise (less uncertain) than either the data or prior alone.*

## 5.5 PRIOR INFORMATION

Given the importance of the prior to Bayesian inference, one important question is where does the prior come from and how do we estimate it? There are many options available, such as information from previous analyses, the literature, meta-analyses, expert opinion, and ecological theory. *The one place the prior* cannot *come from is the data being used in the analysis.* As the name implies, priors should be constructed *prior* to viewing the data being incorporated in the likelihood. Using information from the likelihood to construct the prior, known as "double dipping," leads to falsely overconfident analyses because the same data are used twice, thus artificially inflating the true information content. The other place that priors cannot come from is the posteriors. That is to say, priors should not be altered after the posterior is calculated as a means of "tuning" the analysis. All of us know that it is inappropriate to tune a scientific analysis to get the answer we want, and indeed it is very rare for someone to do this deliberately, but it is very easy for new practitioners to fall into the trap of tweaking prior ranges or variances in hopes of making Bayesian numerical methods converge better or because posterior probability distributions are converging on an assumed prior boundary (that is, the posterior distribution looks like it's smushed up against a wall). In particular, prior ranges should not be readjusted to ranges that are biologically implausible, just to make posteriors more unimodal, as this is a clear indication that a model is trying to get the right answer for the wrong reason. That said, sometimes that error is in our assumptions, not our model structure, so it is fair to realize that our initial thinking was incorrect (for example, assuming a specific process has to have a particular sign) or that we made a units error. New practitioners also frequently want to tweak priors when not enough thought was given to the priors to begin with, for example setting prior uncertainties that may be orders of magnitude beyond what is ecologically realistic. To avoid this, one thing that is perfectly acceptable, and indeed encouraged, when specifying priors is to perform sensitivity and uncertainty analyses (chapter 11) to better understand how specific inputs control model outputs. Based on these analyses one can constrain prior parameters to produce biologically realistic outputs (though what is "biologically realistic" needs to be determined independent of the data). Another option is to subject yourself or colleagues to the formal steps involved in expert elicitation (box 5.2), which forces you to think hard about the values different parameters can take on. While doing so takes time, such effort is usually a valuable investment in understanding your model and refining your hypotheses (that is, thinking carefully about what you expect to see from an analysis).

The statistical literature is full of what are called "uninformative" priors. These are priors chosen to have low information content and minimize the contribution of the prior to the likelihood. On the one hand, this may be appropriate for statisticians, who genuinely have little prior information about ecology and don't know what is an ecologically realistic choice. On the other hand, I feel that uninformative priors are overused by ecologists who somehow feel this is the more objective way of doing statistics. To go back to the neutrino detector example, this is like having no prior opinion about whether the sun just went nova or not. This attempt at objectivity instead results in a greater propensity for spurious results because it is ignoring all previous (objective) data and looking at an individual data set in isolation. Much better, in my opinion, is to take the time to leverage existing knowledge when constructing priors.

The thought process for how to choose a prior probability distribution is very similar to that for choosing a data model. The first question to ask is whether you are starting from scratch or with the posterior from some previous analysis. If the latter, this generally sets the shape and mean of the prior, but it is important to ask whether the previous analysis is actually equivalent to the current one. If so, then the posterior can be used as a prior as is, but if not then it is not uncommon to inflate the variance of the previous posterior to account for this additional uncertainty (for example, data coming from a different site). Another reason for variance inflation would be if previous measurements of a process involved methods that contained little information or may be biased in unknown ways. Next, if you are constructing a prior from scratch based on independent data, literature synthesis, or meta-analysis,

## Box 5.2. Expert Elicitation

Expert elicitation is a commonly used method to estimate probability statements based on expert knowledge. These show up not just in the construction of priors, but also in decision support (chapter 17) when constructing scenarios and alternatives, estimating impacts, and building utility functions. The central challenge with expert elicitation is that humans are not innately wired to think about probabilities and have a wide range of cognitive biases that lead to error and overconfidence (Kahneman 2013). Therefore, an extensive literature has developed on the art and science of trying to elicit judgments. Morgan (2014) provides an excellent synopsis and entry into this literature, while a number of government agencies provide formal guidelines (Ayyub 2000; Boring et al. 2005; Slottje et al. 2008; EPA 2009). Therefore, my first advice on expert elicitation is to not do it without the help of trained social scientists. That said, in many cases experts will be our best, and sometimes only, source of information, and thus we must tread carefully into this domain.

If one proceeds with expert elicitation, Morgan provides the following suggestions:

- Avoid qualitative descriptions of uncertainty (likely, unlikely).
- Develop a formal protocol.
- The mind uses the ease of recovering examples as a proxy for frequency (availability bias). "Help the expert being interviewed to avoid overlooking relevant evidence."
- We tend to *anchor* estimates around an initial guess. Therefore begin elicitation by establishing the highest and lowest possible values.
- Ask follow-up questions to check consistency—for example, what could cause values that are higher/lower than these values.
- Ask about a series of intermediate points to generate a CDF. Examples could include "What's the probability of being greater/less than value *X*?" or "What [upper/lower] value of *X* will have only a 5% chance of being exceeded?" Such elicitations should generally proceed from the extremes inward.
- Ask for the *best* estimate *last* (for example, mean, median).

then that analysis will similarly play a large role in determining the shape of the prior. Constructing priors via meta-analysis is increasingly common and LeBauer et al. (2013) provide a good example of both a Bayesian meta-analysis and the construction of informed priors, while Koricheva et al. (2013) provide an exhaustive treatise on meta-analysis in ecology (see also section 9.1).

Next, consider the case where we are constructing a prior from scratch. When thinking about priors it is pretty rare to encounter a model parameter that is discrete, so my first question is usually whether a parameter has a natural range restriction (for example, nonnegative). Next I think about the shape of the distribution. Do I want something decaying (for example, exponential) or unimodal (for example, Normal)? If unimodal, should it be symmetric or skewed in one direction? Should the distribution decay quickly away from the mean or should it have fat tails? Often at this point I'll have multiple candidate distributions that meet any individual criteria. My next question is whether any one of them is a "natural" fit for the problem, in terms of the definition of the distribution, or whether any distribution will make Bayesian numerical computations more efficient (see section 5.6).

Once the shape of a prior is determined, we need to choose specific values for the prior parameters. For example, in our previous example of estimating a Normal mean, we assumed a $N(\mu_0, \tau)$ prior, but how do we choose $\mu_0$ and $\tau$? If these are not estimated from data, then they need to be chosen carefully, and indeed there is a whole literature on eliciting probability distributions from expert opinion (box 5.2). Such approaches can be used formally to elicit the opinion of external experts, but is also a valuable approach for eliciting your own prior opinions. Given these constraints, which will typically be a list of cumulative probabilities associated with specific parameter values, it is common to construct priors based on *moment matching*. This involves specifying certain summary statistics for the prior and then inverting from this to the prior parameters. For example, you might start with the prior mean, median, or mode, which you think should be centered around a certain value. For mechanistic models that have been inherited from others, this is often the existing "default" parameter. For more statistical models it is not uncommon to center priors around a "no effect" value (for example, setting a regression slope to have a prior mean of zero). Next, you might use a parameter range to define a specific confidence interval (for example, you may be 95% confident that a parameter value falls within a specific range). If the range is not symmetric, that is a good indication that a skewed distribution may be appropriate. Worth noting here is that I did not recommend using a parameter range to define a uniform prior over that range. Except in the case of natural parameter bounds, it is rare that our prior opinion is that all parameters are equally likely, but drops from high probability to zero probability at a discrete threshold.

In defining priors it is very useful both to make plots of the proposed parameters and to draw samples from the distribution. While in theory these two approaches tell you the same thing, in practice it is easy to misinterpret the frequency of extreme parameter draws by looking at the density alone. It is also advisable to test the model using random draws from the prior distributions, as this can illustrate unexpected parameter interactions and whether priors produce outputs in the correct range. The goal here is definitely not to pretune the priors to match the data, but to ensure that priors allow models to explore a wide range of variability without being orders of magnitude outside what is ecologically plausible.

## 5.6 NUMERICAL METHODS FOR BAYES

At this point we have seen only one Bayesian statistical model—fitting a mean with known variance—and in that case the posterior was found by recognizing that it takes on the form of a distribution we already know. So how do we find posteriors more generally? In practice there are few posteriors we can find analytical solutions for because of the need to solve for the normalizing constant in the denominator. For MLE more challenging problems are solved by using numerical optimization, but the same approach does not apply to Bayesian computation because we are not looking for a single "best" parameter value but rather a full distribution. That said,

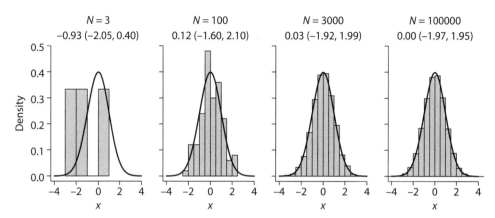

FIGURE 5.6. Approximating a standard Normal density with different numbers of samples ($N$) from a random Normal. As the sample size increases, the histogram becomes a better approximation of the density and the sample mean and 95% quantiles better approximate the true mean and confidence interval: 0.0 (−1.96, 1.96).

any Bayesian method employed does need to find the correct part of parameter space where the posterior distribution is located.

While there are a variety of Bayesian numerical approaches available, most are based on *Monte Carlo* methods (that is, *numerical methods based on repeated random sampling*). Specifically, these methods rely on the same basic sampling concept: *we can approximate any probability distribution with random samples from that distribution*. For example, given a random sample from a distribution we can approximate the PDF using a histogram (figure 5.6). Likewise, any summary statistic that we are interested in can be approximated by the equivalent sample statistic (for example, sample mean, median, standard deviation, quantiles). Furthermore, the numerical accuracy of the histogram and summary statistics is controlled by the number of samples—the more samples you take, the more accurate the approximation. The number of Monte Carlo samples needed in any Bayesian analysis is determined by which summary statistics are required and the desired numerical accuracy of these statistics. As we know from any intro stats course, stable estimates of sample means require fewer samples than are required for stable standard deviations. Similarly, the number of samples required increases as the quantiles we are interested in become more extreme. With $N = 200$ we might get an adequate estimate of a median, since we are balancing 100 samples on each side, but a 95% interval estimate would be determined by the 5 most extreme samples on either side and thus would be pretty coarse. Likewise, if more extreme quantile estimates are needed to address rare events a proportionately larger sample will be needed. Finally, a reminder that this discussion is about the number of Monte Carlo draws required to numerically approximate a posterior, not the number of data points required to run an analysis. Stable numerical estimates can be reached for data sets with very few points, it's just that those intervals will be wide.

Armed with the Monte Carlo sampling concept, how do we use this to approximate a posterior? The key is that *there are various ways to draw random samples from a probability distribution, even if we can't write down an analytical solution for the PDF.* These approaches rely on the ability to calculate the *relative* probability of different parameter values (also known as the *odds ratio*). If we return to the previous example of distinguishing between two alternative hypotheses about $\mu$ given one observation (see figure 5.2), the posterior probability for each would be

$$P(\mu_1|X) = \frac{P(X|\mu_1)P(\mu_1)}{\int P(X|\mu_1)P(\mu_1)}$$

$$P(\mu_2|X) = \frac{P(X|\mu_2)P(\mu_2)}{\int P(X|\mu_2)P(\mu_2)}$$

Then the odds ratio of these posteriors is

$$\frac{P(\mu_1|X)}{P(\mu_2|X)} = \frac{\dfrac{P(X|\mu_1)P(\mu_1)}{P(X)}}{\dfrac{P(X|\mu_2)P(\mu_2)}{P(X)}} = \frac{P(X|\mu_1)P(\mu_1)}{P(X|\mu_2)P(\mu_2)}$$

where the normalizing constant, $P(X)$, cancels out. The relative probabilities thus depend only on the priors and the likelihoods, and we don't need to analytically solve the integral to find the normalizing constant.

Next, let's see how to use these relative probabilities to draw random samples. We'll start by assuming that our current draw is $\mu = \mu_2$ and then propose shifting to $\mu = \mu_1$. To decide whether to do this we draw a random uniform number, $z$, between 0 and 1 and shift to the new value if

$$\frac{P(\mu_1|X)}{P(\mu_2|X)} > z$$

If this *acceptance criteria* is met we shift to $\mu = \mu_1$ and return $\mu_1$ as our random number. If it is not met, then we stay at $\mu = \mu_2$ and return $\mu_2$ as our random number. We then repeat this process again to draw the next random number. If we're still at $\mu = \mu_2$, the process is exactly the same; however, if we switched and are now at $\mu = \mu_1$, then the new acceptance criteria is just the inverse

$$\frac{P(\mu_2|X)}{P(\mu_1|X)} > z$$

In this particular case, $P(\mu_2|X) > P(\mu_1|X)$, so the ratio is >1 and the acceptance criteria is always met. If we repeat this process again and again, it's easy to see that we'll spend more time in the higher probability state but that we don't always stay there. Indeed, the ratio of the number of random samples in each state will converge to $P(\mu_2|X):P(\mu_1|X)$, which is the correct posterior probability.

This algorithm is known as the *Metropolis-Hastings (M-H) algorithm* and can easily be generalized not only to discrete random variables with more than two states but also to continuous random variables, as well as sampling multiple parameters simultaneously. In the general case there is the need for a *jump distribution*, $J(\theta^*|\theta^C)$, which is a probability distribution used to propose new parameter value(s), $\theta^*$, conditional on the current value(s), $\theta^C$. In most cases the acceptance criteria is the same as what we saw before

$$\frac{P(\theta^*|X)}{P(\theta^C|X)} > z$$

However, if the jump distribution is asymmetric, then the acceptance criteria needs to be adjusted because the probability of jumping from $\theta^C$ to $\theta^*$ is not the same as the probability of jumping back

$$\frac{P(\theta^*|X)}{P(\theta^C|X)} \frac{J(\theta^C|\theta^*)}{J(\theta^*|\theta^C)} > z$$

If the jump distribution is symmetric, then this additional term is just 1 and we recover the original criteria. While the M-H algorithm is designed to draw random numbers from a probability distribution it has one property that is similar to the optimization algorithms that are used in numerical MLE—it will always accept a proposed jump that has higher posterior probability, because the acceptance criteria will always be > 1. In this way the algorithm solves not only the normalizing constant problem but also the problem of figuring out where in parameter space the posterior should be located. Indeed, Bayesian numerical methods such as M-H are very robust at searching parameter space; because they always accept "worse" steps with some probability they are less prone to get stuck in local minima than numerical optimization.

The Metropolis-Hastings algorithm is one of a broader class of algorithms collectively referred to as *Markov Chain Monte Carlo (MCMC)*. A *Markov process* is any mathematical or statistical process that has no memory, and thus each transition depends only on the current state of the system. We can see this in M-H because each proposed parameter value depends only on the current parameter value, both in the jump distribution and the acceptance criteria, and not on any parameter values that came before. The effect of this is that a MCMC essentially takes a random walk through the posterior PDF (figure 5.7). However, because random draws from an MCMC do depend on the current value, each draw is not independent of the last, so MCMC draws should be evaluated as a whole distribution, not sequentially.

In addition to M-H there are a wide variety of other MCMC algorithms, such as Gibbs sampling (see box 5.1), as well as other Monte Carlo algorithms, such as rejection sampling (Gilks and Wild 1992) and sequential Monte Carlo (chapter 14), that are beyond the scope of the current discussion. In practice there are a growing number of Bayesian software tools that implement a wide variety of methods "under the hood," so it will be rare that you would have to implement MCMC by hand. That said, understanding the basic theory behind Bayesian numerical methods is critical for understanding and evaluating what you are getting out of software.

## 5.7 EVALUATING MCMC OUTPUT

When performing any statistical analysis there are checks you should perform to ensure that model outputs are sensible and assumptions are met, and the same is true for Bayesian numerical methods such as MCMC. In the following discussion we will continue to use our earlier example of fitting a Normal mean to a single observation, $X = 1.5$. In addition to the checks performed on any statistical model (chapter 16), one of the key checks with MCMC is that the random walk has converged to the correct part of parameter space. One way to check that posteriors have converged is to run multiple replicates (typically 3 to 5) of the MCMC, which are referred to as different *chains*, each starting from different initial conditions. By plotting a *trace plot*, which is the parameter value versus the MCMC sample number, it is often obvious visually whether the different chains have converged to the same part of parameter

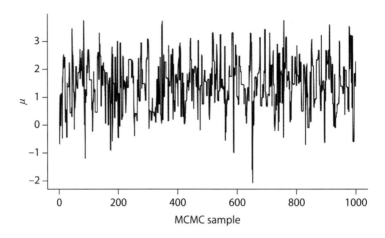

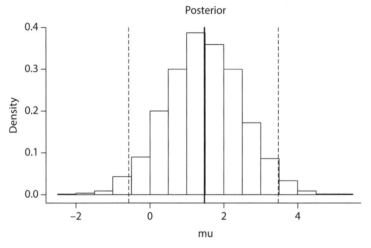

FIGURE 5.7. MCMC output (top panel) for Normal mean with known variance fit to the single observation, $X = 1.5$. In the posterior histogram (bottom panel) vertical lines indicate the position of the mean (solid) and 95% interval (dashed).

space (figure 5.8). In addition to visual inspection, calculating one or more convergence diagnostics is considered good practice as a way of providing some objectivity to assessing convergence. Common diagnostics, such as Gelman-Brooks-Rubin (GBR), Geweke-Brooks, Heidelberger-Welch, and Raftery-Lewis, will be implemented in most Bayesian software (Plummer et al. 2006). GBR, which compares the variance within each MCMC chain to the variance across the chains, is particularly common (figure 5.8). This diagnostic should converge to 1 (equal variance) with most texts recommending a threshold in the range of 1.05–1.10 as indicating convergence. Once the point of convergence has been estimated, the MCMC samples prior to this point, referred to as the *burn-in*, should be discarded. For purposes of generating posterior histograms and summary statistics, samples from all chains can be combined.

Once the burn-in has been removed, it is common to check the amount of *autocorrelation* in the MCMC chains, since as noted earlier sequential samples are not

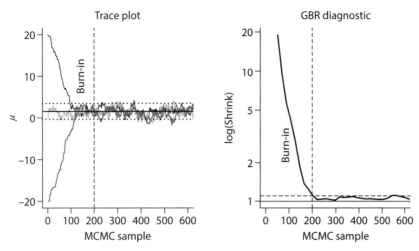

FIGURE 5.8. Convergence diagnostics for three MCMC chains.

independent. Autocorrelation is typically assessed using an *autocorrelation function*, which is a graph of correlation coefficient (on the *y*-axis) versus lag (on the *x*-axis), where lag is the separation between samples. For example, lag 1 autocorrelation is the correlation coefficient between every sample and the previous sample. Lag 0 autocorrelation is always 1, since values are perfectly correlated with themselves. Opinions differ as to whether it is necessary to *thin* (for example, subsample every $k^{th}$ observation) the MCMC output, based on the autocorrelation function, to achieve samples that are approximately independent (Link and Eaton 2012). In either case, it is important to correct summary statistics for autocorrelation when assessing the numerical accuracy of an MCMC estimate. It is common to estimate an autocorrelation-corrected *effective sample size*, when determining whether enough Monte Carlo samples have been taken to meet the goals of an analysis:

$$N_{eff} = N \frac{1 - \rho}{1 + \rho}$$

where $N$ is the original sample size and $\rho$ is the lag 1 autocorrelation.

Once all diagnostics are complete, posterior densities and statistics can be estimated from the MCMC output. It is very important to remember that MCMC gives you the *joint* posterior distribution for all model parameters, not a collection of univariate distributions. This joint distribution embodies not only all the uncertainties about individual parameters but also all the multivariate correlations and nonlinear trade-offs among parameters. *The ability to produce this full joint distribution is an extremely powerful feature of Bayesian methods.* However, it is far too common for these covariances to go unexamined, which can lead to incorrect inferences and predictions. For example, figure 5.9 shows a bivariate joint density with a strong and nonlinear negative correlation among two parameters. In this case the bivariate mode is noticeably offset from the modes of the univariate marginal distributions. Failing to check the bivariate distribution would have led to both a biased estimate of this peak (also known as the *maximum a posteriori probability, MAP*) and an overestimate of the overall uncertainty. Strong covariances among parameters are

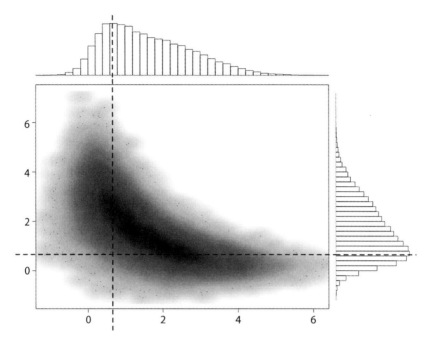

FIGURE 5.9. Bivariate joint posterior density and marginal densities. Dashed lines indicate the mode of each marginal distribution, while the plus sign shows the bivariate mode.

remarkably common, as they can arise both for statistical and ecological reasons when there are trade-offs among parameters. Visualizing trade-offs and covariances among large numbers of parameters can be challenging, but it is important to at least look at the bivariate scatter plots among parameters.

To conclude, this chapter provided a quick introduction to Bayesian concepts and methods. However, it is unlikely that this chapter alone will be sufficient to really dive into Bayesian statistical modeling. The number of books on Bayes for ecologists is increasing rapidly (Clark 2007; Hobbs and Hooten 2015), and beyond this there are a few fairly comprehensive texts in the statistical literature that are less accessible but serve as a valuable reference for specific types of analyses (Gelman et al. 2013).

## 5.8 KEY CONCEPTS

1. The best way to combine models and data is through the use of probabilities.
2. A random variable can take on multiple values, $x_i$, each with probability $p_i$. A probability distribution is a function describing a random variable.
3. The joint probability of $x$ and $y$ is equal to the conditional probability of $x$ given $y$, times the marginal probability of $y$: $P(x,y) = P(x|y)P(y)$.
4. The likelihood expresses the probability of the data, $X$, conditioned on specific values of the model parameters, $\theta$, and thus is commonly referred to as the probability of the data given the model, $P(X|\theta)$.
5. Conceptually, we frequently split the likelihood into the process model, which describes the underlying process, and the data model, which describes the mismatch between the process model and the data.

6. Really understanding where data come from and how they were collected is invaluable when constructing data models.

7. Bayes' theorem allows us to estimate the probability distribution of the model parameters conditioned on the data, $P(\theta|X) = P(X|\theta)P(\theta)/P(X)$, based on the likelihood and the prior probability of $\theta$.

8. The relative impacts of the likelihood and prior on the posterior are determined by their relative uncertainties. Furthermore, the posterior tends to be more precise (less uncertain) than either the data or prior alone.

9. Bayes' theorem provides an inherently iterative approach to inference—the posterior from one analysis can quite naturally become the prior for the next. This formalizes in probability theory how most science is done: new evidence is never considered in isolation but must be viewed in the context of previous data.

10. Priors can be constructed from any data *except* the data going into the likelihood. Ecologists should not shy away from using informative priors—inappropriately broad priors result in a greater propensity for spurious results.

11. Most Bayesian analyses employ Monte Carlo methods (that is, numerical methods based on repeated random sampling) to approximate the posterior distribution with random samples from the distribution.

12. In addition to standard model diagnostics (chapter 16), it is important to check MCMC outputs for convergence, discard the burn-in, and assess parameter correlations.

## 5.9 HANDS-ON ACTIVITIES

https://github.com/EcoForecast/EF_Activities/blob/master/Exercise_05_JAGS.Rmd

- JAGS primer

# 6

## Characterizing Uncertainty

*Synopsis: This chapter dives into the Bayesian methods for characterizing and partitioning sources of error that take us far beyond the classic assumption of independent, homoskedastic (constant variance), and Normal residual error.*

CHAPTER 5 COVERED the basics of Bayesian statistics and highlighted some of the advantages of the Bayesian approach when it comes to forecasting. Key advantages include the ability to estimate a full probability distribution on all parameters (chapter 2), to make use of prior information, and to iteratively update analyses as new data become available, which we will elaborate on in chapters 13 and 14. This framework also makes it easier to combine multiple data sources, which we'll discuss further in chapter 9. Finally, the conditional nature of Bayesian models can be used to build sophisticated analyses from simple parts that can deal with the complexity and uncertainties in real ecological data. This last point—how to characterize the uncertainty in models and data—is the subject of this chapter. In general terms, this chapter is *a primer on how to relax traditional statistical assumptions* about "residual" error—that it is Gaussian, homoskedastic, independent, and represents observation error in the response variable. Similarly, this chapter also introduces approaches for dealing with missing data and for partitioning variability into multiple ecologically meaningful terms. These approaches not only allow for greater flexibility to deal with the complexity of real data, but this realism also leads to better forecasts. Be aware that in this context "better" does not always mean lower predictive uncertainty —many of the approaches described in the following sections can in fact increase uncertainty—but failing to account for many of the factors described here (non-Normal, nonconstant variance, nonindependence, multiple sources of variability and uncertainty) can lead to predictions that are both biased and falsely overconfident. In general it is better to be uncertain than falsely overconfident. Finally, the approaches introduced here serve as important building blocks for more advanced chapters.

### 6.1 NON-GAUSSIAN ERROR

One of the most common challenges with ecological data is that they frequently do not conform to the standard assumption of Gaussian error. In dealing with non-Gaussian data, I generally eschew the laundry list of transformations laid out in introductory statistics, as they are frequently abused and used out of convenience rather than a reasoned assessment of the data in hand. Just because a transformation makes a data set look more Normal (for example, an arctan transform of proportion data), doesn't mean that it offers a sensible interpretation. Furthermore, the effects of *Jensen's inequality (that the mean of a nonlinear function is not equal to the function evaluated at the mean of its inputs, $\overline{f(x)} \neq f(\overline{x})$)* are routinely ignored when results are back-transformed, which leads to biased projections. Even beyond this, in simulation experiments (where the true parameters are known) data transformations consistently perform worse than choosing an appropriate non-Normal distribution (O'Hara and Kotze 2010). Fortunately, there is a seemingly endless list of other probability distributions available for use besides the Normal (table 6.1). Developing a conceptual familiarity with the most common distributions is important for the type of statistical modeling discussed in this book—*they are the essential building blocks for build data models and priors.*

So how does one go about choosing between different probability distributions (figure 6.1)? As discussed in chapter 5, a good starting point is to consider the constraints imposed by the nature of the data themselves. For example, are the data continuous or discrete? Can the data take on negative values? Is there some other hard upper or lower bound on the data, such as probabilities that must be bound between zero and one? In addition, while some distributions are chosen because they provide a good fit to the data, the choice between multiple similarly shaped distributions is often informed by an understanding of the processes that different distributions represent. For example, the Negative Binomial distribution is a discrete, nonnegative distribution that represents the count of the number of "failures" that occur before you observe $n$ "successes" (for example, how many times you have to flip a coin to get a total of $n$ tails). However, the Negative Binomial also arises as a generalization of the Poisson distribution where the rate parameter, $\lambda$, follows a Gamma distribution. Thus, the Negative Binomial is often used to describe data that seem to come from a Poisson process but have a higher variance than the Poisson. Likewise, many distributions are closely related to one another, often occurring as generalizations or special cases of other distributions. For example, the Geometric distribution is a special case of the Negative Binomial where $n = 1$ (for example, the distribution of the number of coin tosses until you get a tails [Stoppard 1994]).

In addition to using distributions "off-the-shelf," it is possible to combine distributions in various ways to construct data models that better describe the complexities of real data. One way to do this is to have the parameters of one distribution vary according to another distribution, such as the Negative Binomial example earlier. Similarly, the Student's t distribution arises from a Normal distribution where the precision (1/variance) is drawn from a Gamma distribution, which is used to account for the uncertainty in the variance. Alternatively, we can combine distributions as additive mixtures. For example, a combination of Normals could be used to describe a multimodal distribution where each peak gets a different weight, $w_i$

$$P(x) = w_1 N(\mu_1, \sigma_1^2) + w_2 N(\mu_2, \sigma_2^2) + \cdots + w_n N(\mu_n, \sigma_n^2)$$

where the weights have to sum to 1 (figure 6.2, top). Similarly, a mixture of exponentials might describe decomposition data, where compounds that differ in recalcitrance would be described with different decay rates, or survival data, where different groups are exposed to different hazards. This approach can also be used to combine different distributions. One common application of this is the generation of "zero-inflated" distributions, such as the zero-inflated Poisson (ZIP), which addresses

TABLE 6.1. Common Probability Distributions and Their Characteristics

| Name | Equation | Bound | Interpretations and Relation to Other Distributions |
|------|----------|-------|-----------------------------------------------------|
| *Discrete* | | | |
| Bernoulli($p$) | $p^x(1-p)^{1-x}$ | $[0,1]$ | Success ($x = 1$) with probability $p$. Binomial with $n = 1$. |
| Binomial($n,p$) | $\binom{n}{x}p^x(1-p)^{n-x}$ | $[0 \dots N]$ | Number of successes, with probability $p$, given $n$ trials. |
| Poisson($\lambda$) | $\frac{\lambda^x}{x!}e^{-\lambda}$ | $0,\infty$ | Number of events, occurring at rate $\lambda$, to occur over a fixed interval. |
| Negative Binomial($n,p$) | $\binom{x+n-1}{x}p^n(1-p)^x$ | $0,\infty$ | Number of trials, with probability $p$, before $n$ successes occur. Also a Poisson-Gamma mixture. |
| Geometric($p$) | $p(1-p)^x$ | $0,\infty$ | Number of trials needed before a success occurs. Special case of negative binomial with $n = 1$. |
| *Continuous* | | | |
| Uniform($a,b$) | $\frac{1}{b-a}$ | $a,b$ | All values between $a$ and $b$ have equal probability. |
| Beta($\alpha,\beta$) | $\frac{\Gamma(\alpha+\beta)}{\Gamma(\alpha)\Gamma(\beta)}x^{\alpha-1}(1-x)^{\beta-1}$ | $0,1$ | Probability of success given $\alpha$ out of $\alpha + \beta$ trials were successful. Beta(1,1) = Unif(0,1). |
| Exponential($\lambda$) | $\lambda e^{-\lambda x}$ | $0,\infty$ | The interval between events in a Poisson process at rate $\lambda$. |
| Laplace($\mu,b$) | $\frac{1}{2b}\exp\left(-\frac{|x-\mu|}{b}\right)$ | $-\infty,\infty$ | Two-sided exponential. |
| Weibull($\lambda,k$) | $\frac{k}{\lambda}\left(\frac{x}{\lambda}\right)^{k-1}e^{-\left(\frac{x}{\lambda}\right)^k}$ | $0,\infty$ | Generalization of exponential, where $\lambda$ changes according to $k$. Weibull($\lambda$,1) = Exp($\lambda$). |
| Gamma($a,r$) | $\frac{r^a}{\Gamma(a)}x^{a-1}e^{-rx}$ | $0,\infty$ | Sum of $a$ Exp($r$) variables, Gamma(1,$r$) = Exp($r$). As a prior precision, Gamma(*sample size*/2,*sum of squares*/2). Gamma($n$/2,2) = $\chi^2(n)$. |
| Normal($\mu, \sigma$) | $\frac{1}{\sqrt{2\pi}\sigma}\exp\left(-\frac{(x-\mu)^2}{2\sigma^2}\right)$ | $-\infty,\infty$ | Maximum entropy distribution given a mean and variance. |
| Lognormal($\mu,\sigma$) | $\frac{1}{\sqrt{2\pi}\sigma x}\exp\left(-\frac{(\log x-\mu)^2}{2\sigma^2}\right)$ | $0,\infty$ | $\log(x)$ is Normally distributed. Note the mean is $e^{\mu+\sigma^2/2}$ not $\mu$. |
| Student's t($n$) | $\frac{\Gamma\left(\frac{n+1}{2}\right)}{\sqrt{\pi n}\Gamma\left(\frac{n}{2}\right)}(1+x^{2/n})^{-\frac{n+1}{2}}$ | $-\infty,\infty$ | Normal with a Gamma distributed precision. |
| Cauchy($\mu,\gamma$) | $\frac{1}{\pi\gamma\left[1+\left(\frac{x-\mu}{\gamma}\right)^2\right]}$ | $-\infty,\infty$ | Ratio of two Normal variables. Has no mean and variance. |
| Chi-squared($n$) | $\frac{1}{2^{\frac{n}{2}}\Gamma\left(\frac{n}{2}\right)}x^{\frac{n}{2}-1}e^{-\frac{x}{2}}$ | $0,\infty$ | Sum of squares of $n$ standard Normal variables. |

*Note*: All of the equations are parameterized with $x$ as the random variable.

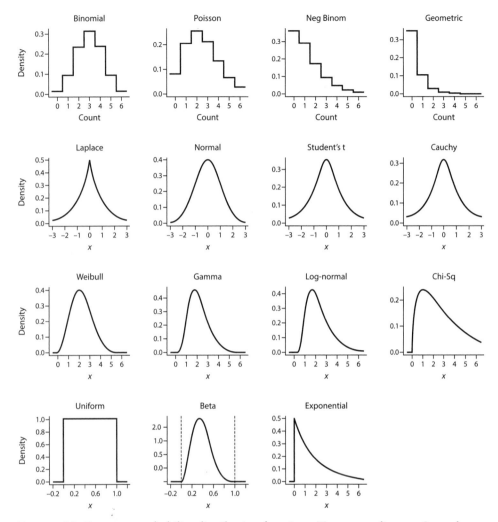

FIGURE 6.1. Common probability distribution functions. Top row = discrete. Second row = continuous symmetric. Third row + Exponential = continuous and bound at zero. Bottom row (Uniform and Beta) = continuous with upper and lower bounds.

the observation that in many data sets zeros are much more common than would be expected by chance (figure 6.2, bottom):

$$ZIP(\lambda,\theta) = \begin{cases} x = 0 & \underset{absent}{\theta} + \underbrace{(1-\theta)e^{-\lambda}}_{unobserved} \\ x > 0 & \underbrace{(1-\theta)\,\frac{\lambda^x e^{-\lambda}}{x!}}_{observed} \end{cases}$$

For example, with count data from a biological survey, $\theta$ would be the probability that the species is absent, and $(1 - \theta)e^{-\lambda}$ would be the standard Poisson odds of not observing a species that is present with abundance $\lambda$, conditional on the species being present $(1 - \theta)$ (Wenger and Freeman 2008). All of the observed presences are likewise

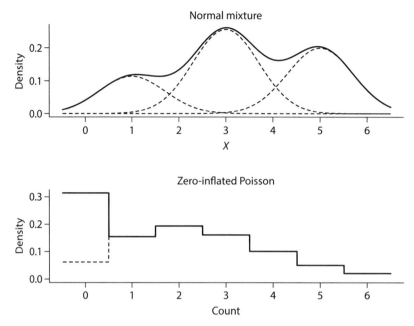

FIGURE 6.2. Mixture distributions. Top: Mixture of three Normal distributions. Bottom: Zero-inflated Poisson with 25% probability of absence. Dashed line indicates the Poisson expectation for a zero observation given presence.

expressed conditional on the probability of being present. What is important to note here is that because there are two ways in which a species can have a zero count (true absence versus unobserved), the full ZIP model has to be fit all at once—the probability of absence can't just be estimated from the proportion of absences in the data. Zero-inflation can also be applied to continuous distributions, such as the log-normal, where zeros are not allowed but are sometimes recorded in data. Zero-inflation is a much more defensible alternative than the common practice of adding an arbitrary value to all zero data, a procedure known to produce biased results. It is important to note that the weights given to the different distributions in a mixture can be modeled themselves. For example, in the ZIP the weight, $\theta$, is the probability of presence versus absence, which might be further modeled as a logistic of environmental variables that affect presence. Furthermore, there might be a different set of covariates used to model $\lambda$, the abundance of the species conditioned on its presence.

A final way that distributions are frequently modified is through truncation. This means that, for one or both tails of a distribution, a boundary is imposed outside of which the distribution is assumed to have zero probability. When truncating a distribution the remainder of the distribution then has to be renormalized to ensure that the probability distribution integrates to one. In general, such bounds should be specified a priori based on some firm conceptual understanding of the process being studied, rather than being estimated from data (for example, highest or lowest value observed to date). It is important to remember that truncation ascribes zero probability to observing a value outside the bounds, a statement that is both hard to justify and estimate from data alone. It is also important to be aware that truncated distributions can sometimes produce unexpected results and can be difficult to fit. For example, in a truncated Normal distribution the mean and variance are no longer equivalent to the parameters $\mu$ and $\sigma^2$. Indeed, a Normal truncated at zero can have a negative $\mu$, which will produce a distribution that just corresponds to the upper tail of the Normal distribution. Furthermore, this new distribution, which will look roughly exponential, can be very hard to fit, as decreasing the mean but increasing the variance will lead to a distribution that looks very similar.

So how, in practical terms, does one use a non-Gaussian distribution? To fit data directly to a distribution is generally straightforward, as one can apply standard likelihood or Bayesian numerical methods to estimate the parameters of the distribution. Also, it is not difficult to use a non-Gaussian distributions as priors. What is often more challenging is to construct a data model where the process model controls the parameters in the distribution. If we take linear regression as an example, there we're assuming that the residuals are Normally distributed. However, this model can easily be written in terms of the error distribution itself and a process model that controls a parameter in that model—in this case the mean, $\mu$

$$y_i \sim N(\mu_i, \sigma^2)$$

$$\mu_i = \beta_0 + \beta_1 x_i$$

In other words, the linear process model is predicting $\mu_i$, the expected value of $y_i$ for every $x_i$, and in the data model the actual $y_i$ are assumed to be Normally distributed around these predictions. However, some distributions, such as the Exponential, Binomial, and Poisson, have a mean parameter but the range of possible $y$ values is bounded over some range. For these distributions a linear model is often a poor choice because it can produce invalid predictions (for example, negative means being passed to zero-bound distributions). The traditional solution to this problem, which serves as the basis for *generalized linear models* (GLMs), is the use of a link function, which converts a real number to one in the domain of the distribution's mean. While the link function is often thought of as a transformation, it is important to realize that the link function is actually a core part of the process model. For example, using the same notation as earlier, we can write out one of the most common GLMs, logistic regression, as

$$y_i = Bern(p_i) \qquad \text{Data model}$$

$$p_i = logit(\theta_i) = \frac{1}{1 + e^{-\theta_i}} \quad \text{Link function} \qquad\qquad (6.1)$$

$$\theta_i = \beta_0 + \beta_1 x_i \qquad \text{Linear model}$$

In a logistic regression the data are Boolean (TRUE/FALSE, 0/1) and thus follow a Bernoulli distribution that describes the probability of the observed event occurring. Because probabilities have to be bounded between 0 and 1, we model this using a sigmoidal function called the logit. Since the one parameter of the logit function, $\theta$, can take on any real number, we can further model $\theta$ as a linear function of any covariates we are interested in. Similar link functions exist for other distributions, such as the Poisson and Exponential, that can take on only positive values, with the Poisson GLM being a common choice for count data. GLMs can even be generalized to multivariate distributions, such as the Multinomial GLM, which generalizes the Binomial (logistic regression) to when the response data ($y$) involves multiple categories rather than just two. However, what is critical to remember is that in all these cases your process model is no longer the linear model, but instead is the combination of the link function and linear model. *There is nothing fundamentally wrong with this, but it is critical to be aware of the assumptions made in a GLM and to be sure that the process model used is the one you intended.* More generally, there's nothing that says that one has to use a standard link function, or that the process model has to contain a linear model at its core. Any function that conforms to the bounds of the distribution you are using can serve as the process model.

The other challenge that frequently occurs when using non-Gaussian distributions is that in many distributions the parameters in the model do not correspond to a mean and variance, like they do in the Normal. For example, the Gamma distribution has a mean of $a/r$ and a variance of $a/r^2$. In such cases one option is to construct process models for the parameters in the distribution. This is particularly sensible if the conceptual understanding of the parameters in the distribution corresponds to how you are modeling your process. For the Gamma, that might mean modeling the number of events, $a$, and their rate, $r$. Alternatively, when using the Gamma to model a precision, that might correspond to the sample size, $a = n/2$, and sum of squares, $r = SS/2$. However, if the Gamma was chosen because it provided a good fit to non-negative data, then this isn't as straightforward. In this case one might specify the process model in terms of the *moments* of the distribution, and then transform from the moments to the distribution's parameters. For example, for Gamma process models, the mean and variance can be transformed back to $a$ and $r$ as $r = \mu/\sigma^2$ and $a = \mu r$.

Overall, this section demonstrates that there is a wide range of options for probability distributions that can be used in data models to describe error and as prior distributions. As we'll see later in this chapter these distributions can also be used to partition many other sources of uncertainty and variability.

## ▨ 6.2 HETEROSKEDASTICITY

Another common challenge that arises with real data is *heteroskedasticity*, which is when *the observed variability around the mean is not constant.* Not only does a changing variance violate the assumptions of most classic statistical models, but it will have large impacts on our forecasts, leading to overconfidence in some conditions and underconfidence in others (figure 6.3). One of the first things to consider when encountering heteroskedasticity is whether the probability distribution you are using is already heteroskedastic, and if not whether you should switch to a distribution that is. For example, in the lognormal, Poisson, and Negative Binomial the variance naturally increases as the mean increases. The second option is to model the variance explicitly, just as we modeled the mean in section 6.1. To consider a simple regression example, if exploratory data analysis suggests that the standard deviation of the residuals is increasing linearly with the covariate $x$, then we could write this as

$$y_i \sim N(\mu_i, \sigma_i^2)$$

$$\mu_i = \beta_0 + \beta_1 x_i$$

$$\sigma_i = \alpha_0 + \alpha_1 x_i$$

Such a model can then be fit using standard maximum likelihood or Bayesian methods (figure 6.3). The process model for the standard deviation need not be linear, but

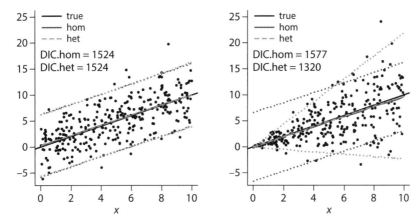

FIGURE 6.3. Impact of heteroskedasticity on model fit. Left: For a homoskedastic data set a model that accounts for heteroskedasticity produces virtually the same fit. Right: For a heteroskedastic data set the model that accounts for heteroskedasticity has a much better model fit (lower deviance information criteria [DIC]) and a more realistic interval estimate. The homoskedastic model consistently overestimates uncertainty at low $x$ and underestimates uncertainty at high $x$. DIC is a Bayesian model selection metric analogous to the Akaike information criterion [AIC] (Spiegelhalter et al. 2002).

could be any process model that captures both our conceptual understanding of the process and the observed variability. As with developing a model for the mean, the keys to developing a good model that describes the variance (or any other higher moments) are ample exploratory data analysis and diagnostics, combined with critical thinking about the processes that generate variability.

## 6.3 OBSERVATION ERROR

A discussion on modeling observation error is straightforward if we believe that all of the observed residual error is measurement error and that there is no uncertainty in the model structure, covariates, or drivers. However, there is often an appreciable amount of true variability and heterogeneity in ecological systems (chapter 2). In distinguishing observation error from process error, I separate the problem into two cases, observation error in the response variable(s), $y$, versus error in the covariates or drivers, $x$. As with other problems, it is simplest to explain these concepts in the context of a simple regression model, though the approaches transfer directly to more sophisticated models. Let's start with errors in $y$. If we take $y_i$ to be the true value of the response variable and $y_{i,obs}$ to be the observed value, then we might model this as

$$y_{i,obs} \sim g(y_i) \qquad \text{Data model}$$

$$y_i = \beta_0 + \beta_1 x + \varepsilon_{i,add} \qquad \text{Process model}$$

$$\varepsilon_{i,add} \sim N(0, \tau_{add}) \qquad \text{Process error}$$

where $g(y_i)$ is some probabilistic function describing the relationship between the true and observed values of $y$. The data model, $g(y_i)$, could be as simple as Gaussian noise, $y_{i,obs} \sim N(y_i, \tau_{obs})$, or could be a sophisticated representation of the observation process.

For example, in remote sensing the data model may involve a series of complex operations related to instrument calibrations, atmospheric corrections, geolocation, topographic corrections, view-angle corrections, and so on (Townshend et al. 2012).

A common challenge in modeling residual error is separating observation and process error. If we write out the simple example earlier assuming Gaussian noise

$$y_{i,obs} = \beta_0 + \beta_1 x + \boldsymbol{\varepsilon}_{i,add} + \boldsymbol{\delta}_{i,obs}$$

$$\varepsilon_{i,add} \sim N(0, \tau_{add})$$

$$\delta_{i,obs} \sim N(0, \tau_{obs})$$

we can see that there is very little basis for partitioning the residual error (shown in bold) into observation versus process error, as $\varepsilon$ and $\delta$ can take on any value so long as they sum to the observed residual. This problem is known as *nonidentifiability*.

There are a few ways to improve the ability to partition uncertainty. The first, and most obvious, is if there is external information available about the observation error, $\tau_{obs}$, such as calibration information for an instrument. In this case this information can be included as an informative prior. Here the observations are unlikely to provide much new information about the observation error, but specifying an informative prior on the observation error serves to better identify the process error. Similarly, in some cases the sampling design itself generates a well-defined sampling error (for example, Poisson count data) that can be built into the data model. The second option is if there is QA/QC information internal to the data set, such as measurements made on the same unit using different instruments or made by different observers. This will increase the identifiability of parameters even if remeasurement is done on only a subset of the data. (Section 11.4 will touch on methods used to determine what fraction of the data would need to be remeasured.) The key to being able to distinguish process and observation error, in these two cases, isn't that the observation error has to be small, but that our uncertainty about the observation error needs to be small (that is, that $\tau_{obs}$ is well constrained).

The third way that process and observation error can be partitioned is if the data are collected in such a way that observation error and process error lead to different predictions. One way to do this is to take advantage of the fact that observation error does not propagate forward in the model but process error does. Examples of this in time and in space will be the subject of chapter 8. Another way to do so is to rely on structure in the process error, such that it does not apply independently to every single measurement but may instead be associated with observation units that are observed repeatedly, such as individual organisms, plots, and so on. Similarly, the observation error may likewise be structured, such as differences in precision among observers or instruments. Accounting for such structure is one of the uses of hierarchical models, the topic of section 6.5.

Let us now return to the second form of observation error, that associated with the covariates, drivers, or inputs to the process model. In the statistics literature, this is often referred to as the *errors in variables (EIV)* problem. To give an example, consider the linear model:

$$y_i \sim N(\mu_i, \sigma_i^2)$$

$$\mu_i = \beta_0 + \beta_1 x_i$$

$$x_{i,obs} \sim g_{EIV}(x_i)$$

As with observation error on the response, $g_{EIV}(x)$ can be as simple as Gaussian noise or can be very detailed. For example, if errors in a data set were reported on an individual measurement basis, such as a standard deviation $\tau_i$, this could be modeled as $x_{i,obs} \sim N(x_i, \tau_i^2)$.

Another common, though often underappreciated, example of EIV is the simple problem of instrument calibration. Let's say we're trying to understand the relationship between plant growth, which we'll record in terms of change in height of the $i^{th}$ plant, $\Delta H_i$, and volumetric soil moisture, $\theta_i$. We might assume this relationship to be an asymptotic function, such as the Monod

$$E[\Delta H_i] = G_{max} \frac{\theta_i}{\theta_i + k}$$

where $G_{max}$ is the maximum growth rate, and $k$ is the half-saturation constant, the soil moisture level where the plant grows at $G_{max}/2$. Let's next assume that the residual error in height growth is Normal, and for the moment let's not worry about the distinction between observation and process error in the $y$:

$$\Delta H_i \sim N(E[\Delta H_i], \sigma^2)$$

However, let's assume that we don't want to measure the soil moisture content by drying a soil core, as this would quickly turn our plots to Swiss cheese, but instead rely on an indirect proxy such as time domain reflectometry (TDR), which uses two electrical probes to measure soil impedance. Next, let's assume that we diligently collected paired measurements of TDR and soil moisture in order to calibrate our instrument to the local soils. So now our true $x$ is soil moisture, $\theta$, our $x_{obs}$ is TDR, and the data model is now the calibration curve (figure 6.4), which we'll take to be

$$\theta_i \sim N(\alpha_0 + \alpha_1 TDR_i, \tau_x)$$

It is worth noting that the soil moisture in the growth model is now a random variable that comes from the *predictive* distribution of the calibration curve. That means that it includes both the observation error $\tau_x$ and the parameter error in $\alpha$, and while the latter will decrease asymptotically as the size of the calibration data set increases, the observation error will not. Indeed, the distribution of the errors in the $\theta_i$ need not be equal, and they would not be in the preceding example because of the hourglass shape to the calibration curve's predictive interval. Taken as a whole, despite the fact that our calibration curve has a high $R^2$, failing to propagate the errors in calibration results in the growth/moisture relationship being overconfident and biased (CI excludes the true curve by a substantial margin).

Unlike errors in $y$, identifiability tends not to be quite so difficult with uncertainty in the $x$. However, covariance between the errors in the $x$ and $y$ directions would indicate that these parameters are poorly constrained and trading-off between one another. Still, as with observation error in the response variable, the best option is to have independent calibration data or QA/QC data with which to constrain the estimates of these variances. Indeed, if the errors in inputs are well characterized (even if they are not small), one could skip the problem of trying to estimate the input errors and instead just focus on propagating those errors into the model outputs (section 11.2).

Finally, an increasingly common challenge, for both $x$ and $y$ variables, is accounting for the uncertainties associated with *derived data products*, rather than raw

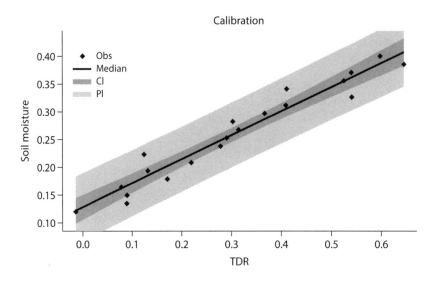

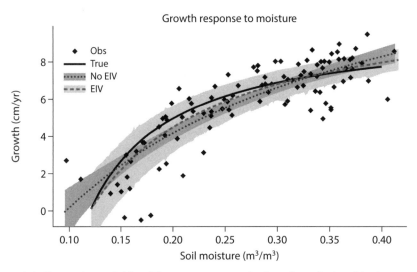

FIGURE 6.4. Errors in variables. The upper curve depicts the relationship between the variable of interest, soil moisture, and the proxy variable, TDR, including the 95% confidence and predictive intervals for the linear model between the two. This relationship has an $r^2$ of approximately 95%. The lower panel depicts the relationship between growth and soil moisture. The model with EIV treats soil moisture as a probability distribution, while the model without EIV assumes that the $x$ values take on the mean value from the calibration curve. All data are simulated from a known "true" growth curve and calibration curve, which allows us to identify that the model without EIV is both biased and overconfident.

observations. For example, an ecological model may use a gridded meteorological data product, such as the North American Regional Reanalysis (Messinger et al. 2006), that is itself the output of a data-model fusion scheme between a numerical weather model and a whole suite of ground, airborne, and satellite measurements. Likewise, a large fraction of the remote sensing used by ecologists is derived data (for example, land cover, leaf area) rather than raw observations. A number of significant challenges exist with the use of derived data products. First, data products come in many shades of gray between products that are very close to raw observations and those that are predominantly model outputs, and while it is clear that a model should be fit to data, not another model, where to draw the line on which data products are OK to treat as data and which are not is often unclear. Likewise, because data products come with built-in assumptions that might not hold everywhere, there is a greater risk of introducing systematic errors associated with unknown or uncorrected biases. Second, many derived data products do not come with uncertainty estimates. Though the trend for reporting uncertainty seems to be improving, it can be challenging to track down what sources of uncertainty are and are not included in a reported uncertainty estimate. Just because a standard deviation is reported doesn't automatically mean that standard error is the appropriate one to use. For example, an estimate of variability in space or time is not an observation error, but rather process variability, and thus should not be used as a prior on observation error. A third and related challenge is understanding and accurately incorporating the sample size of a derived data product. For example, if I measure 100 points in space and then interpolate those observations onto a $1000 \times 1000$ grid, $n = 100$ not 1,000,000. As a general conclusion, failing to include and propagate the uncertainties in a data product can seriously reduce their usefulness and, if unaccounted for, may render analyses based on these products falsely overconfident about a potentially wrong prediction. As we saw in chapter 5, the results of a Bayesian analysis are often weighted by their uncertainties, indicating that uncertainties in data and models are critical to get right, as they often determine the outcome of our analyses and forecasts. This is not to say that data products should never be used, but rather that they need to be used with caution.

## 6.4 MISSING DATA AND INVERSE MODELING

Most ecologists have a painful familiarity with the missing data problem: the failure of an instrument that required you to throw out a chunk of data on a covariate; the day a lightning storm caused you to call the day early after having measured only a subset of individuals; the random errors in field notes that can't be teased out (was that measurement 6.8 cm or 68 cm?). After years of hard work on a project, you begin your analysis only to find that there are small, random holes in the data for unavoidable reasons. At first this doesn't seem like a big problem, but then, as you move onto more complex, multivariate analyses, you begin to realize that the union of all these small holes in different covariates adds up to a nontrivial fraction of data lost. Whole rows of hard-won data end up excluded due to the lack of individual covariates.

While Stats 101 leaves little recourse for the missing data problem, it turns out that the Bayesian approach of treating unknowns as random variables (that is, as probability distributions) provides a way to move forward—integrate over all possible values that the missing data could take. This process of estimating missing values is known as *imputing*, while the process of averaging over a MCMC sample of values is referred to as *multiple imputation*, since you are filling the gap in the data with multiple alternative values rather than just interpolating a single value.

As with all Bayesian analyses, the process of multiple imputation starts with specifying a prior, in this case the prior on the missing data. Applied to our standard example of linear regression, this might look like

$$y_i \sim N(\beta_0 + \beta_1 x_i, \sigma_i^2) \quad \text{Data and process models}$$

$$x_{miss} \sim N(\mu_x, \tau_x) \quad \text{Missing data model}$$

$$\mu_x \sim N(\mu_{x0}, \tau_{x0}) \quad \text{Prior mean}$$

$$1/\tau_x \sim Gamma(a_x, r_x) \quad \text{Prior variance}$$

where $x_{miss}$ is the subset of $x$ that are missing, and $\mu_x$ and $\tau_x$ are the mean and variance of the $x$ data, which are informed by the observed (nonmissing) $x$'s. If we look at the conditionals for estimating $x_{miss}$, they involve both the regression and the missing data models

$$x_{miss}|y \propto N(y|\beta_0 + \beta_1 x_{miss}, \sigma_i^2) N(x_{miss}|\mu_x, \tau_x)$$

which means that the missing $x$ is informed by both the distribution of all other $x$'s and the observation for this particular $y$.

Figure 6.5 shows an example of a missing data model being used to estimate the probability distribution of $x_{miss}$ conditional on an observation of $y = 7.5$. In this

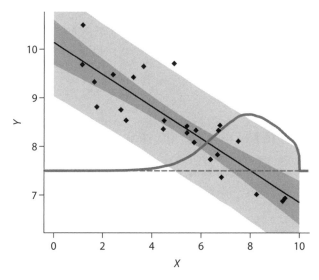

FIGURE 6.5. Missing data linear model. The black straight line and gray confidence and predictive intervals depict the linear regression fit to the observed data (black diamonds). The horizontal dashed line indicates the observed $y$ that is paired with the missing $x$ observation, while the thick curved line indicates the posterior probability distribution for the value of the missing $x$ (vertical dimension in units of probability density). This distribution is consistent with the horizontal span of the regression predictive interval.

pseudo-data example the $x$ data were, by design, restricted to between 0 and 10, so I chose to use Unif(0,10) as the missing data prior. This decision highlights the point made in section 6.1 that we are in no way locked into Gaussian assumptions and should *choose distributions appropriate to the problem*. What we observe in this example is that the posterior probability distribution for $x_{miss}$ is clearly updated by the observation of $y$ and takes on a nonstandard distribution. This distribution nonetheless corresponds to what we would expect based on the predictive interval (light gray).

The missing data model, which uses an observation of the *state variable y* to infer the distribution of the *state variable x*, is more generally an example of a simple state-variable data assimilation approach, *Bayesian inverse modeling*. In the more general case we can use an existing model for the relationship between $x$ and $y$, combined with the prior for $x$, to estimate some new unobserved $x_{new}$ conditional on new observations of the response, $y_{new}$. This approach has many uses in data assimilation and forecasting when we want to make inferences about $x$ rather than $y$. For example, we might have a model that predicts the reflectance spectra of a body of water ($y$) given a set of traits related to algal density ($x$). We can then use remotely sensed observations of the reflectance ($y_{new}$) to make inferences about the state of the algal community ($x_{new}$). The posterior of $x_{new}$ might be directly of interest, may feed into an assessment of hypoxia or algal impacts on fisheries, or become the initial condition for a model predicting the growth and spread of an algal bloom. Finally, Bayes' theorem plays a critical role in inverse modeling because it allows us to formally generate an estimate of $p(x_{new}|y_{new})$ given a model for $p(y|x)$ and a prior expectation on $p(x)$.

The astute observer may have noted that, in the preceding discussion of missing data and inversions, there is no information provided by the missing $x$ in the fitting of the model itself. Indeed, you get the same results (figure 6.5) whether the missing datum is included or not. However, the situation changes once we move to a model with multiple covariates, such as multiple regression. Imagine we have $k$ covariates and in row $i$ we are missing the first covariate, $x_{i1}$, but observe all the others, $x_{i2} \ldots x_{ik}$. In this case our estimate of $x_{i1}$ in the missing data model is conditional on $y$ and $x_{i2} \ldots x_{ik}$; therefore, our estimate of $x_{i1}$ is much more informative. More importantly, we are now able to include a whole row of data, $x_{i2} \ldots x_{ik}$, that we previously had to drop. It is this ability, *to include data that previously would have been dropped*, that increases our inferential power.

One final but critical note: missing data models can only be used when the data are *missing at random*. Formally, this means that the probability that a datum is observed cannot depend on the missing value

$$p(observation|x,x_{miss},\theta) = P(observation|x,\theta)$$

where $\theta$ is the set of parameters in the model. Note that by this definition, the missing data can depend upon the model and other data, but missingness cannot be systematic. The previous examples in this section are all instances of data that are missing at random. As another example, if samples at night are more likely to be missing, as is common with eddy-covariance data (Moffat et al. 2007), we can't use the distribution of all observed values to estimate the missing data model for the nighttime data because they are systematically different from the daytime values. Furthermore, in many cases when data are not missing at random we can't even take the traditional route of excluding missing data. For example, in eddy-covariance you can't compute the mean carbon flux just from the sample mean of the data without accounting for the missing data because the sample is biased toward daytime values. However, if we *condition on time of day* (which is allowable because it's in *x*), we can use the nighttime values in a missing data model provided that the mechanism causing these values to be absent is not related to the process under study. However, if missing data are missing systematically (for example, missing nighttime fluxes are different from evenings with observations) then they cannot be imputed.

## ▬▬ 6.5 HIERARCHICAL MODELS AND PROCESS ERROR

A common challenge in ecological data analysis and forecasting is to correctly partition different sources of variability. Imagine you have data from a collection of random plots, within which you collect multiple samples, $x_{p,i}$, where $p$ is plot and $i$ is an individual sample. Let's start by ignoring covariates and just fit a mean to the data. One option is to pool all the data across plots and to calculate a single mean and variance, $\mu$ and $\sigma^2$ (figure 6.6). However, pooling data ignores the across-plot variability and treats individuals within a plot as independent. This option highlights another common challenge with ecological data, the lack of independence of observations within measurement units. Another option is to assume that all plots are different and to calculate independent means and variances for each plot, $\mu_p$ and $\sigma_p^2$. Fitting means independently implies that if we were to add another plot to the system that we couldn't use *any* information from the existing plots to predict the new plot. This occurs because of our assumption of independence—in this analysis there's no probability distribution spanning across plots that we can use to make a prediction. Furthermore, assuming there are $k$ plots, the number of parameters goes up from 2 to $2k$, which causes the parameter error to increase considerably (for example, if $k = 100$, we go from 2 parameters to 200). At the same time the residual error will go down, since the differences among plots will explain some of the variability. Which of these two options leads to the lowest overall uncertainty is a classic model selection problem and will be determined by the sample sizes, number of plots, and their variability. However, neither case is really a good description for what's going on in the system, since in one case we're ignoring all plot-to-plot variability, while in the other we're assuming that the plots are completely unrelated, neither of which is likely correct.

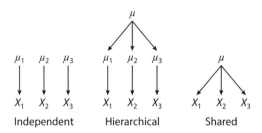

FIGURE 6.6. Hierarchical models represent the continuum of possibilities between treating data sets as independent versus treating them as identical. As a result, they partition process variability between the different levels of the hierarchy.

Instead of choosing between these options, let's think of these two models as end-points in a continuum from truly identical plots to truly independent plots. What we would like is the capacity to model that continuum. This is the goal of hierarchical models. Statistically speaking, the hierarchy in hierarchical models is with respect to the *parameters* in the model, not the data. Therefore, *hierarchical models are designed to capture structured variability (not error) in a model's parameters*. In this simple example (figure 6.6, middle), a hierarchical model would have two levels reflecting within-plot versus across-plot variability. At the plot level we are fitting parameters to each plot—in this case, just a plot-level mean—but at the next level up we are writing a model that describes how the plots relate to one another—in this case, a probability distribution capturing both the across-plot mean *and the across-plot variability*. An example of this, assuming Normal means and Gamma precisions, would be

$$x_{p,i} \sim N(\mu_p, \sigma^2) \qquad \text{Within-plot model}$$

$$\mu_p \sim N(\mu, \tau) \qquad \text{Across-plot model}$$

$$\mu \sim N(\mu_0, \tau_0) \qquad \text{Global mean prior}$$

$$1/\sigma^2 \sim Gamma(a_\sigma, r_\sigma) \qquad \text{Within-plot variance prior}$$

$$1/\tau \sim Gamma(a_\tau, r_\tau) \qquad \text{Among-plot variance prior}$$

The immediate outcome of this model is that it partitions the overall variability into two terms, one describing the within-plot variability in $x$, $\sigma^2$, and the other, $\tau$, describing across-plot variability in $\mu$. More generally, hierarchical models provide a means for partitioning the sources and scales of variability in a process rather than lumping all sources of variability into a residual. In other words, they are a powerful tool for *estimating and partitioning process error*. In many cases hierarchical models provide a more accurate model of the processes we are studying and explicitly acknowledge that different processes operate at different scales. Furthermore, they provide a mechanism for *accounting* for variability in a process or scale even if we cannot yet *explain* that variability in our process model.

The other strength of hierarchical models is that they can be used to account for the lack of independence among measurement units. In our example we have accounted for the lack of independence of individual measurements within the plot, since the deviation of the plot as a whole from the overall mean is accounted for by $\mu_p$. In a hierarchical model, plots (or any other unit of measure) are not modeled as independent from one another, nor are they modeled as identical. Indeed, while the number of individual parameters estimated in a hierarchical model might actually go up compared to the independent fits, since the $\mu_p$ being estimated are not independent, *the effective number of parameters lies between the two extremes*. Furthermore, the position of the model along this continuum is determined by the data. The more similar the plots are to one another, the smaller the among plot variance, and the smaller the effective sample size ends up being, because the $\mu_p$ become less and less independent.

Hierarchical models don't have to consider just one source of process error, but can accommodate multiple sources of variability, such as differences among years or even individual-level differences (Clark et al. 2007). In deciding how to structure hierarchical models, the general rules of thumb from more traditional *random effects* models tend to be a useful starting point: *an effect is random if it would be different if the study was replicated* (for example, individuals, plots, watersheds, years, and so on). In other words, they are *the things that we might choose to replicate over*, rather than the things we would include as a covariate. Mathematically, random effects are a subclass of hierarchical models that estimate how parameters deviate from their higher-level expectation. For example, in figure 6.6 we model each of the plot-level means as distributed around a global mean, $\mu_p \sim N(\mu,\tau)$. The random effects model would instead write the process model as $\mu_p = \mu + \alpha_p$, where $\alpha_p$ is the plot-level deviation from the global mean. $\alpha_p$ would then be modeled hierarchically with a mean of 0 (that is, unbiased), $\alpha_p \sim N(0,\tau)$, with a prior on $\tau$ as before. One strength of the random effects framework is that it makes it easier to write models that include multiple sources of variability. For example, if the plots we described earlier were also measured over multiple years, then our data varies both by plot ($p$) and by year ($y$), and we could write this down as $\mu_{p,y} = \mu + \alpha_p + \alpha_y$, with hierarchical models on both $\alpha_p$ and $\alpha_y$.

Hierarchical models can be further generalized beyond just random effects to include *fixed effects* (that is, covariates) as part of the process models at different scales to explain the different scales and sources of variability (Dietze et al. 2008). A model that combines fixed and random effects is generally referred to as a *mixed effects* model. One useful general approach is to begin by modeling the sources of variability in an ecological process as random effects, and then as analysis proceeds, or as more data are accumulated, to begin to "chip away" at the random effects with progressively more detailed process models focused on the largest sources of variability. In this way the initial random effects model *accounts* for different sources of variability and the incremental addition of fixed effects aims to *explain* that variability. However, the random effects will generally still be retained until the fixed effects explain all the variability in the process. The other advantage of the mixed effects framework is that the magnitudes of the random effects provide guidance about the what scales need explaining. In our previous example with random effects for plot and year, if the year variance is much larger than the plot variance that points us toward a different set of potential explanatory variables than if the plot variance dominates the year variance.

An important subset of mixed models is the class of models known as *generalized linear mixed models (GLMM)*, which combines generalized linear models (section 6.1), which are linear models connected to non-Gaussian distributions via nonlinear link functions, with hierarchical mixed effects, to produce a broad and flexible set of parametric models. For example, logistic regression (equation 6.1) could be generalized to

$$y = Bern(p) \qquad \text{Data model}$$

$$p = logit(\theta) = \frac{1}{1 + e^{-\theta}} \qquad \text{Link function}$$

$$\theta = \beta_0 + \beta_1 x_1 + \beta_2 x_2 + \alpha_1 + \alpha_2 \qquad \text{Mixed effects model}$$

$$\alpha_i \sim N(0,\tau_i) \qquad \text{Random effect}$$

$$1/\tau_i \sim Gamma(a_i,r_i) \qquad \text{Random effect variance}$$

where there are multiple fixed effects $(x_j)$ and random effects $(\alpha_i)$. Random effects are neither the same for all rows of data (like $\beta$) nor different for every row, but generally vary according to one or more sampling variables (for example, $\alpha_1 = $ year, $\alpha_2 = $ block, and so on). The random effect variance quantifies the variability across such groupings (for example, year-to-year variability). More broadly, hierarchical and random effects models can be applied to parameters in all types of process models, not just linear models, such as the example from section 2.3.4, where the logistic growth model's intrinsic growth rate and carrying capacity varied from year to year.

Since hierarchical models are capturing process error, that error propagates into forecasts (chapter 2). For example, consider the previous example of a collection of plots that have been measured over multiple years. In this case we used a random effects model to partition both plot-to-plot and year-to-year variability. Our prediction for one of these established plots in the next year would account for the persistent, idiosyncratic differences of that individual plot estimated from previous years, but would have to integrate over the uncertainty associated with a new year (figure 6.7, "Year"). In other words, $\alpha_p$ would be known (with uncertainty), but we'd have to use $\alpha_y \sim N(0,\tau_y)$ to sample over the possible values of $\alpha_y$. By contrast, a prediction for a

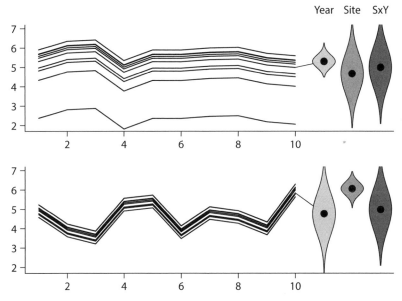

FIGURE 6.7. Impact of random effects on forecasts. Both panels show the trajectories of 10 sites over 10 years and have the same total variance. In the top panel 75% of the variability is site-to-site, while in the bottom panel 75% is year-to-year. On the right-hand side the violin plots under "Year" depict the prediction for a previously known site for the next new year. In this case the top example (low year-to-year variability) is more predictable. "Site" depicts the prediction for a new site in year 10, and in this case the bottom example (low site-to-site variability) is more predictable. Finally, "SxY" depicts the prediction for a new site in a new year, which is identical for both panels since they have the same total variability.

new location in the current year would have a known year effect, $\alpha_y$, but would have to integrate over the across-plot variability, $\alpha_p$ (figure 6.7, "Site"). The prediction for a new plot and new year would have the highest uncertainty because it would have to integrate over both year, $\alpha_y$, and plot, $\alpha_p$, effects (figure 6.7, "SxY"). All of this may sound obvious, but it is very different from how prediction would be made if process error were treated as residual error.

## 6.6 AUTOCORRELATION

In the final section of this chapter, I want to briefly touch on more traditional models for dealing with the lack of independence of observations, which are based on including autocorrelation. Typically this is handled within the error covariance matrix. In the Gaussian data models we've discussed so far there is typically a single constant variance. An equivalent way of writing such a model would be as a multivariate Gaussian with a covariance matrix that is zero everywhere except the diagonal, and that diagonal just repeats that same constant variance, with one row for every observation. The heteroskedastic models discussed in section 6.2 could similarly be represented as a diagonal covariance matrix, but with the variances varying from row to row. The key conceptual idea here is to account for the lack of independence of observations by filling in the off-diagonal elements of the covariance matrix by modeling the correlations among observations. (Recall from chapter 2, any covariance matrix, $\Sigma$, can be broken down into a diagonal matrix of standard deviations, $D$, and a matrix of correlation coefficients, $C$ according to the formula $\Sigma = DCD$.)

While accounting for correlation through the covariance matrix is a perfectly sensible thing to do, the challenge is that there are a huge number of parameters to estimate in an unconstrained covariance matrix. Imagine that you have five observations that are related to one other in time or in space. The covariance matrix would thus be $5 \times 5$, and even assuming that the matrix is symmetric (that is, that the correlation between point i and point j is the same as between j and i), that gives $n(n + 1)/2 = 15$ parameters, including the diagonal variance term. The typical solution to this problem is to assume that the correlation between any two points, in space or time, is a decreasing function of *distance*:

$$Cov[x_i, x_j] = \sigma^2 \cdot f(\|s_i - s_j\|)$$

where $\sigma^2$ is the variance, $f$ is the correlation function, and $x_i$ is the *value* of the system at location $s_i$. The term $\|s_i - s_j\|$ indicates that we are taking some scalar measure of distance between these two locations, though this does not have to be a Euclidean distance. This structure makes two important assumptions. The first is *second-order stationarity*, which means that correlation is just a function of distance (that is, that it doesn't depend upon location) and that there is no trend in the residuals. There can be trends in the underlying data, but the process model needs to explain that trend. The second assumption is *isotropy*, which means that the correlation structure is the same in all directions. Given these assumptions the problem becomes much more tractable. For example, if we assume that the correlation decreases exponentially with distance

$$f(\|s_i - s_j\|) = \rho^{\|s_i - s_j\|}$$

then the covariance matrix becomes a two-parameter model ($\sigma^2$ and $\rho$). Here $\rho$ is the autocorrelation coefficient, which is bound between $-1$ and $1$. If we assume that all the pairwise distances between points are stored in a matrix, $M_D$, then the solution for the full covariance matrix becomes

$$\Sigma = \frac{\sigma^2}{1-\rho}\,\rho^{M_D} \tag{6.2}$$

Other autocorrelation functions are possible besides the exponential, such as the Gaussian and the Matérn. Indeed, any function can serve as an autocorrelation function provided that it leads to a positive definite covariance matrix. However, the proof that a given function is positive definite is nontrivial so most analyses use well-established autocorrelation functions. While most frequently used to deal with autocorrelation in time or space, this approach is general to any situation where the degree of autocorrelation is related to some measure of *distance*, such as phylogenetic distance or community similarity.

The fact that autocorrelation models are based on correlation matrices means that applications of this approach are generally limited to models using Gaussian likelihoods (which is not true for the alternative state-space approach; see chapter 8). However, despite the intimidation factor that many ecologists face when encountering spatial and temporal autocorrelation models, they are really not that much harder to write than other models. For example, if we have a standard linear regression model, we can write the likelihood out in matrix form as

$$\vec{y} \sim N(X\vec{\beta}, \sigma^2 I)$$

where $y$ is the vector of observations, $X$ is the matrix of covariates, $\beta$ is the vector of regression coefficients, $\sigma^2$ is the variance, and $I$ is the identity matrix (a diagonal matrix of just 1's). The covariance matrix in this case ends up having $\sigma^2$ on the diagonal (the covariance of an observation with itself is just its variance) and zero covariance everywhere else. The equivalent regression model taking autocorrelation into account is just

$$\vec{y} \sim N(X\vec{\beta}, \Sigma)$$

where $\Sigma$ is the covariance matrix given by equation 6.2. This approach can also be used to account for autocorrelation associated with repeated measures of the same experimental units, the only difference being that in that case you have a $\vec{y}_i$ for each of many units, but units presumably share the same variance and autocorrelation parameters (or these are structured according to some other hierarchical model). It is also possible to combine autocorrelated error with other sources of error. For example, the following model has an additional additive variance $\tau$:

$$\vec{y} \sim N(X\vec{\beta}, \Sigma + \tau I)$$

which might be used to account for observation error or some other form of process error not accounted for by the spatial or temporal process. This is particularly necessary when you have observations that are very close to one another because the autocorrelation model predicts that as two observations approach each other they should become identical (as the distance goes to 0, the correlation goes to 1). In the case of spatial models, this additional variance is typically referred to as the "nugget."

Prediction with an autocorrelation model is a bit more complicated than for a traditional model because now you need to create a new distance matrix that includes all of the original observations and all of the new points being predicted. For spatial problems, predicting new locations with a Gaussian autocorrelation model is known as *Kriging* and is a well-established approach in spatial statistics.

Readers interested in a more detailed exploration of these topics should consult statistical textbooks on Hierarchical Bayesian methods such as (Gelman et al. 2013). From time-series and spatial statistics there is also an extensive literature on how to work with models with distance-based covariance matrices (Diggle 1990; Banerjee et al. 2003; Cressie and Wikle 2011).

## 6.7 KEY CONCEPTS

1. This chapter is a primer on how to relax traditional statistical assumptions about "residual" error—that it is Gaussian, homoskedastic, independent, and represents observation error in the response variable.
2. The mean of a nonlinear function is not equal to the function evaluated at the mean of its inputs, $\overline{f(x)} \neq f(\overline{x})$ (Jensen's inequality).
3. It is better to thoughtfully choose the appropriate data model than transform data to meet Normality assumptions.
4. Nonconstant variance (heteroskedasticity) can be explicitly modeled parametrically.
5. Separating observation error from processes errors requires informative priors, internal QA/QC data, or for errors to be structured (for example, in time, space, or hierarchical).
6. Ignoring errors in variables (covariates, inputs) leads to falsely overconfident and potentially biased conclusions.
7. Derived data products should be used with caution, especially if they lack a rigorous partitioning of uncertainties.
8. Missing data can be accounted for, with uncertainty, via multiple imputation provided data are missing at random.
9. Bayesian inversion allows us to formally generate an estimate of $p(x_{new}|y_{new})$ given a model for $p(y|x)$. Such inversions can depend strongly on the prior expectation on $p(x)$.
10. Hierarchical models represent a continuum between treating groupings (years, plots, and so on) as identical versus independent (completely unrelated). In doing so they account for, and quantitatively partition, within-versus across-group variability, potentially for multiple factors simultaneously. These approaches can be extended to represent variability model parameters.
11. Partitioning process errors improves forecasts because different errors do not propagate the same ways in space and time.
12. Autocorrelation models assume that the correlation between any two points in space or time is a decreasing function of distance

## 6.8 HANDS-ON ACTIVITIES

https://github.com/EcoForecast/EF_Activities

- Probability distributions and mixtures
- Modeling heteroskedasticity
- Errors in variables: sensor calibration
- Missing data
- Random effects in forecasts
- Adding covariances to models

# 7

# Case Study: Biodiversity, Populations, and Endangered Species

*Synopsis: Population ecology has a long history of quantitative modeling and provided some of the earliest examples of sophisticated Bayesian analyses in ecology. Have these insights translated into improved predictions of impacts on populations (for example, endangered species)? How can we translate these successes to the community level (biodiversity)?*

## 7.1 ENDANGERED SPECIES

Population ecology has a deep tradition of sophisticated population modeling (Lotka 1910; Volterra 1926; Turchin 2003). However, as argued with the logistic growth example in chapter 2, the development and teaching of such models has at times focused more on theory than linking models to data to make explicit quantitative predictions. There are obviously exceptions to this, and arguably the largest and most important of these has been the assessment of the viability of threatened and endangered species (Morris et al. 1999; Morris and Doak 2002). The introduction of legal protections both domestically (for example, US Endangered Species Act, 1973) and internationally (International Union for Conservation of Nature [IUCN] Red List, 1964; Convention on International Trade in Endangered Species of Wild Fauna and Flora [CITES], 1973) created a strong demand for quantitative assessments of endangered populations. This in turn generated a requirement not just to assess population size, but to forecast population trajectories and assess the efficacy of proposed management scenarios. That said, it has also generated a tremendous amount of litigation, which reinforces the need for transparency and adherence to best practices (chapters 3 and 4).

Rather than making explicit forecasts of endangered population sizes through time, endangered species forecasting (and population forecasting more generally) has traditionally focused on the *intrinsic rate of increase, r* (Birch 1948), and the *asymptotic growth rate, $\lambda$* (Leslie 1945), both of which are used to infer whether a population is declining ($\lambda < 1$ or $r < 0$) in the long term (that is, asymptotically). The standard tool for population viability analysis has long been the age- or stage-structured matrix model (Caswell 2001). This model typically takes on a form similar to:

$$
\begin{bmatrix} n_1 \\ n_2 \\ \vdots \\ n_k \end{bmatrix}_{t+1} = \begin{bmatrix} s_1 + f_1 & f_2 & \cdots & f_k \\ g_1 & s_2 & & \\ & \ddots & \ddots & \\ & & g_k & s_k \end{bmatrix} \begin{bmatrix} n_1 \\ n_2 \\ \vdots \\ n_k \end{bmatrix}_t
\tag{7.1}
$$

where $n_{i,t}$ is the population size ($n$) of stage or age $i$ at time $t$. In this example the species has been divided into a total of $k$ age or stage classes and the elements in the matrix are the class-specific fecundity, $f$, in the first row, the probability of survival within a class, $s$, along the diagonal and the probability of growing to the next class, $g$, along the off-diagonal. More generally, in stage-structured models it is not uncommon for transitions to appear in other parts of the matrix (for example, a reproductive individual can switch back to a nonreproductive state), while in an age-structured model (for example, Leslie matrix), $s = 0$ because you can't both survive and stay the same age.

The method for solving for $\lambda$, which mathematically turns out to be the dominant eigenvalue of the matrix, is well-established in the literature, as are a plethora of variants of matrix models that introduce environmental covariates on $f$, $s$, and $g$ or add stochasticity to these parameters (Caswell 2001). Also well established are sensitivity analysis approaches (chapter 11) for determining which transitions have the largest impact on $\lambda$. These analyses are often used to evaluate alternative management scenarios. For example, the demonstrated high sensitivity of $\lambda$ to adult survival in loggerhead sea turtles (*Caretta caretta*) played a key role in the introduction of Turtle Exclusion Devices (TEDs) to commercial fishing nets (Crouse and Crowder 1987; Crowder et al. 1994). Similarly, early matrix population models of northern spotted owls (*Strix occidentalis caurina*) demonstrated the sensitivity of the species to logging impacts (Lande 1988), which eventually led to 80% reductions in timber harvesting in the Pacific Northwest and pushed the spotted owl into the national spotlight. Sensitivity analysis approaches such as these have also been extended to the analysis of population transients, rather than just the asymptotic trajectory, which can be particularly important for species subject to frequent disturbance or otherwise far from their stable age distribution (Fox and Gurevitch 2000). More recently, integral projection models (IPMs) have gained popularity as a continuous-stage alternative to discrete matrix models (Ellner and Rees 2006, 2007; Merow et al. 2014). The strength and flexibility of IPMs comes from the use of parametric approaches (for example, generalized linear mixed models) to model vital demographic rates as continuous functions of size/stage (as well as other covariates), but ultimately the analyses and tools are very similar to matrix models, and everything discussed here extends to IPMs. Readers interested in IPMs are encouraged to study the review by Merow et al. (2014), which has an exhaustive 184-page appendix of examples and code.

One important intellectual development in population forecasting has been a deeper appreciation of variability, and the distinction between uncertainty and variability. The propagation of uncertainty in model parameters into the uncertainty in $\lambda$ has long been part of matrix modeling, but parameter estimates often either lumped all data together (see figure 6.6, Shared) or were generated by fitting completely separate models for different years or different populations as a way of incorporating variability (see figure 6.6, Independent). The former approach leads to more tightly constrained parameter estimates, but at the expense of ignoring real-world variability and heterogeneity. The latter accommodates this variability but usually at the expense of greatly increased parameter uncertainty and a questionable ability to extrapolate. As discussed in section 6.5, hierarchical models provide a means of finding a middle ground between these two extremes by modeling variability (for example, across years or among subpopulations) as being drawn from some shared distribution rather than being completely independent. This allows parameter estimates to

FIGURE 7.1. Florida scrub mint (*Dicerandra frutescens*). Photo credit: Reed Bowman.

share information, which is critical as it is not uncommon for certain demographic parameters to be missing for specific years or sites. It also allows forecasts to formally distinguish between sources of uncertainty, which are reducible through data collection, and sources of variability, which are irreducible and need to be accommodated in management plans.

As a case study on the impact of these distinctions, consider the endangered Florida scrub mint, *Dicerandra frutescens ssp. frutescens (Lamiaceae)* (figure 7.1). This species is native to Florida oak–hickory scrub and sand pine scrub communities that are periodically top-killed by fire, but which resprout vigorously. These communities contain many fire-dependent species that are endemic to the region and threatened by a combination of fire suppression and land use change—in that context species like the scrub mint are indicators of threats to an entire ecoregion. The scrub mint itself is a short-statured, short-lived (<8 years) perennial that is killed by fire but dependent upon the canopy openings created by fire. To assess the viability of this species, Evans et al. (2010) constructed a stage-structured matrix model (equation 7.1) using 20 years of demographic data from five populations that varied in their fire history. Within the matrix model they then modeled the effect of time since fire (TSF) on each demographic rate. To account for the variability in demographic rates, submodels for the component vital rates included a common random year effect (YEAR) that is shared across all stages, a stage-specific random year effect ($\sigma_{yr}^{STAGE}$), and a stage-specific random population effect ($\sigma_{pop}^{STAGE}$) (figure 7.2). Because the impact of the shared year effect potentially varies in magnitude and direction among stages, a stage-specific coefficient, $\beta_2^{STAGE}$, rescales the YEAR effect for every stage, except the reference one (in this case, the number of flowering branches on large flowering plants). Importantly, rather than just introducing uncorrelated random noise, this model accommodates both the known (time since fire) and unknown (year, year × stage, population) sources of population variability while accounting for the shared covariance across stages. To capture these covariances, the entire model needs to be fit at once, rather than fitting parameters individually. Furthermore, it's worth noting that traditional bootstrapping approaches to uncertainty estimation would not account for this process variability and could create unrealistic parameter combinations by combining data across years or populations.

The model itself consists of a set of eleven generalized linear mixed models (sections 6.1 and 6.5) fit simultaneously and connected to each other via a standard stage-structured matrix population model (equation 7.1). Each component model involves a linear model of the aforementioned fixed (TSF) and random (YEAR, yr, pop) effects; since there was only one fixed effect, no interaction terms were considered. Starting from reproduction, both the number of flowering branches per plant and the number of seeds per branch were modeled using a Poisson data model and a log link. Separate regressions were fit for the number of branches/plant for each of three reproductive stages (small flowering, medium flowering, large flowering), but the seeds per branch regression was shared across stages. Next, seed germination and seedling survival are both Bernoulli processes (0/1), so they were modeled with a Binomial data model and logit link (also known as logistic regression). It's worth noting that there were no direct observations of seedbank densities and few direct observations of seed production so the seedbank stage of the model was completely latent. Finally, the model had five multinomial regressions for the transitions out of the five nonseed stages (seedling, vegetative, small flowering, medium flowering, large flowering). For each stage, all four transitions were modeled (reversion to seedlings was not allowed), though for seedlings direct transitions to reproductive were rare.

The analysis by Evans et al. (2010) demonstrated that in Florida scrub mint the probability of extinction was lowest when the fire return interval was 24 to 30 years. This frequency occurred through the balance of reproduction (germination, probability of being reproductive, number of flowering plants), which declines with time since fire, and survival, which declines with time since fire in young plants but increases in large plants. The analysis also showed that even with 20 years of data available, over half the uncertainty in the population growth rate was due to parameter uncertainty rather than either deterministic (time since fire) or random process variability. Furthermore, while a fire interval of 24 to 30 years was found to be best, even at this interval the average population is still in decline because the negative effects on fecundity and recruitment are greater than the positive effects on survival. Evans et al. speculate that this decline may be due to changes in climate or megafauna, but neither factor was explicitly in the analysis.

One thing that is striking about the Florida scrub mint example is the amount of uncertainty in population growth rate (figure 7.2). If we know that populations in the real world are variable in space and time, then traditional approaches to population viability analysis are overconfident. This overconfidence can lead managers to make

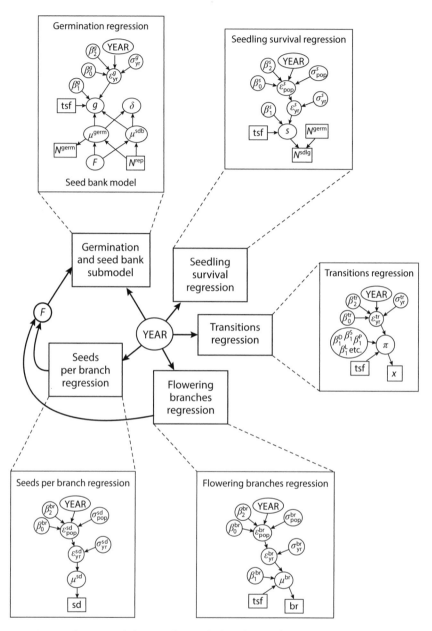

FIGURE 7.2. Population viability analysis of Florida scrub mint. Overall model structure describing the life cycle of the species (center) and directed graphs for each submodel. Submodels included a fixed effect for time since fire (TSF), a common random year effect (YEAR) shared across all stages, a stage-specific random year effect ($\sigma_{yr}^{STAGE}$), and a stage-specific random population effect ($\sigma_{pop}^{STAGE}$). Fit parameters in each submodel are an intercept ($\beta_0$), slope on TSF ($\beta_1$), scaling factor on YEAR ($\beta_2$). Data for each submodel are depicted by boxes, while the remaining additional parameters in circles (for example, $\mu$, $\varepsilon$) are latent variables showing the hierarchical structure.

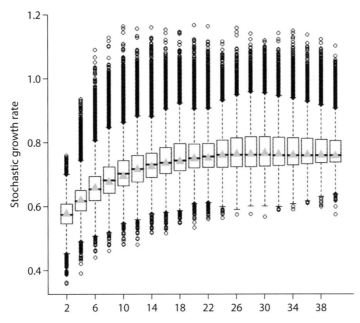

FIGURE 7.2. (*continued*) Simulation of population growth rate as a function of fire return interval.

poor choices (see chapter 17), and in the long run degrades the credibility of the ecological forecaster. In addition, looking at $\lambda$ at a population level hides the variability among subpopulations—some of which are likely doing far worse than the traditional model suggests, while others are likely doing better than average and may actually be sustainable. Indeed, observational data suggest that the nondeclining subpopulations in older stands are taking advantage of persistent canopy gaps, which illustrates the importance of accounting for random effects across subpopulations.

In a more general sense, where population forecasters are succeeding at ecological forecasting is in accommodating the complexities of real-world data (chapter 6). They have long been accustomed to non-Gaussian and heteroskedastic data, as most population data are counts. They are pushing beyond variability in space and time and diving into individual-level demographic variability (de Valpine et al. 2014). So where are the opportunities and frontiers? The first is in data fusion (chapter 9). On one hand, matrix and IPM models already exhibit a fusion of data sources, as different observations are frequently used to constrain different vital rates (fecundity, growth, survival). However, most population models are built either top-down using population-level time-series data or assembled piecemeal from the bottom up using demographic data, but not the fusion of the two. Integrating the two is important because the bottom-up is frequently more data rich but can miss covariances, trade-offs, and cross-scale interactions (Clark et al. 2014).

Another opportunity is in the development of automated forecasts that continuously integrate new observations. Developing automated forecasts requires the integration of informatic (chapters 3 and 4) and statistical (chapters 13 and 14) workflows, but could provide faster turnaround for already overworked agencies. Automatically updated forecasts could more quickly identify emerging challenges for species, and could focus additional measurements when and where they are most needed—for example, to differentiate natural fluctuations from the crossing of thresholds. Indeed, Evans et al. conclude their analysis of Florida scrub mint by advocating for population viability analysis to occur in an ecological forecasting framework, whereby new data are continually assimilated into the model and forecasts are updated based on explicit exogenous scenarios (fire in this case).

## 7.2 BIODIVERSITY

In moving from endangered species to biodiversity, we switch from population to community ecology and encounter a considerably wider range of modeling approaches used for forecasting. The simplest models of species diversity are classic extensions to the logistic, such as Lotka-Volterra, or simple resource competition models, such as the $R^*$ models (Tilman 1982); however, these are rarely used to make quantitative forecasts. As Botkin et al. (2010) note when discussing attempts to forecast biodiversity responses to global warming, "theoretical models of these effects are limited … and should not be taken literally."

In practice, models used for biodiversity forecasting range from simple species-area curves (Preston 1962) to complex semi-mechanistic models based on detailed demography and/or ecophysiology. Twenty years ago, when introducing one such semi-mechanistic model (SORTIE), Pacala et al. (1996) succinctly summarized the challenges of predicting biodiversity. The core problem is that in a network of $n$ interacting species the number of direct pairwise interactions increases as $n^2$ while the number of indirect feedbacks increases as $n!$, which is clearly intractable given that there are millions of species on the planet. However, if all interactions matter equally in a system, then none of them would matter much (average importance proportional to $1/n!$). While purists in complex systems theory may argue the point, my experience has been that most systems are dominated by a modest number of strong interactions that explain most of the variability. Examples of this are keystone species and ecosystem engineers.

That said, there is growing evidence that simple models that reduce the complexity of the world to a small number of factors are not sufficient. These "low dimensional" approaches may function as a useful first pass, but it is important that subsequent iterations move forward, which may require categorical changes in approach. For example, a considerable amount of effort has gone into constructing species distribution models (also known as *climate envelope models*) as a means of forecasting the future range of species under climate change scenarios. The basic concept is to build a statistical model relating a species' current range to environmental variables (for example, temperature, precipitation). Future scenarios for those predictor variables are then used to project the range map into the future. This is a useful first-order approximation, but fundamentally these correlative models fail to accommodate basic ecological constraints (Ibáñez et al. 2006; Evans et al. 2016), such as:

- how these same environmental variables impact an organism's physiology (internal state) and demography (growth, mortality, reproduction, dispersal);
- biophysical, biogeochemical, stoichiometric, historical, and evolutionary constraints;
- species interactions and the difference between a fundamental and realized niche; and

- landscape- to regional-scale processes, such as disturbance, fragmentation, and management.

For example, I argued earlier that communities may be dominated by a small number of interactions, but climate envelopes don't include any such interactions. Furthermore, climate envelopes don't necessarily validate well against historical and paleoecological data (Williams et al. 2013).

Refining biodiversity forecasts requires that we seek approaches that accommodate the complexities of the real world without becoming mired in them. To do so we need to clarify the (multiple) driver variables we're interested in and the spatial, temporal, and taxonomic scales relevant to the forecast. For example, at a global, millennial, and coarse-taxonomic scale, simple life zone classifications explain a considerable amount of the observed variability in biodiversity and may be sufficient. However, the same approach (that is, climate envelopes) will likely perform progressively worse as we zoom into finer scales, because other ecological factors will become steadily more important. Similarly, we might use very different models if, instead of climate, we want to understand the impacts of extreme weather events, $CO_2$ fertilization, land use, invasive species, management, or pollutants. Unfortunately, all too often we need to address all of these factors simultaneously. However, as noted earlier, we know that ultimately all these factors will not be equally important in forecasting species responses. Focusing on the dominant drivers of change can help simplify both monitoring efforts and forecast models. The risk of this is that by relying on any one set of constraints to simplify the problem, we may be missing important factors outside those constraints. These missing factors are the "unknown unknowns" that too often paralyze community ecologists and keep us trapped in case studies and skeptical of forecasts. There is no magic bullet for dealing with the challenges of complexity. There will always be a risk of unexpected outcomes, and that risk needs to be openly acknowledged and discussed. However, risk should not prevent us from making forecasts altogether, as the need for scientific input in the area of biodiversity is pressing, and the process of making testable predictions is an important part of the learning process.

The Catch-22 in modeling the dominant processes in a system is that too often we feel we have to add complexity to a model (for example, adding a new process) to identify which processes are important. In other words, we take an educated guess about what's wrong with the model, and then only after we've gone through the effort of coding, calibrating, and validating the new model do we discover whether we've improved the model. This approach highlights two needs in model development. First is the need to accommodate and partition the uncertainties associated with factors that are ignored (chapter 6). Second is the need for a more evidence-based approach to model refinement that focuses added complexity where it's most needed. Addressing both these needs requires a quantitative identification of the dominant scales (in space, time, taxonomy, and so on) associated with variability in a process (section 6.5). This approach is in contrast with the "case study" approach of demonstrating that some factor is "important" locally without first checking whether that's the relevant scale to be working on. For example, plant ecologists have spent decades studying how local competitive interactions among neighbors affects growth (Bugmann 2001), implicitly under the hypothesis that most variability in growth occurs at the local scale. At the other extreme, global vegetation modelers have studied how climate drives growth (Cramer et al. 2001), implicitly under the hypothesis that most variability in growth is at the regional scale. However neither stopped to check how much variability in growth was associated with these two scales (Mantooth et al. in prep.). It's entirely plausible that the local interactions that look so important when you're standing next to a tree may be responsible for a tiny percentage of the overall variability in growth. It's also plausible that gradients in regional-scale mean growth may be dwarfed by emergent local interactions. Finally, it's also plausible that some other scale is driving the variability and both were missing the real action. My point isn't to argue which hypothesis is right, but to point out that both sides (myself included!) were wrong to spend decades adding complexity to models without first stopping to figure out what scale they should be working on.

Practically, hierarchical models provide a way to accommodate the variability that's occurring at different scales even if we cannot yet explain that variability. As mentioned in section 6.5, a potentially useful approach is to start with a simple null process model (for example, $f(x) = \beta_0$) but with a rich partitioning of variability across scales (for example, using random effects). Estimating the variability associated with different scales helps focus additional complexity specifically at the scales that matter most. For example, if regional-scale variability is largest, model development should focus on regional-scale drivers (for example, climate). As determinism is added at a specific scale, this will reduce the unexplained "random-effect" variability associated with that scale. Attention may eventually shift to processes at another scale, once that scale comes to dominate the unexplained variability. If approached iteratively, this strategy allows us to tackle the largest sources of uncertainty in the order of their importance. There are two important nuances to remember about this approach. First, as processes are added to explain variation at any particular scale, don't drop the random effect at that scale since the deterministic portion will never explain 100% of the variability. Second, don't forget about the residual! If, after partitioning out observation errors, the scale-independent residual process error is the dominant source of variability, then the previous approach is just chipping away at minor components. In this case, if the process variability is ecologically important but not associated with any known spatial, temporal, or phylogenetic scale, we need to reassess our basic concepts. Your system may be driven by some latent process you haven't uncovered yet or is simply inherently unpredictable (for example, chaotic, irreducibly stochastic, or computationally irreducible; see section 2.5).

## 7.3 KEY CONCEPTS

1. Age- and stage-structured models (matrix models, IPMs) are a common framework for making population projections.
2. Hierarchical models can accommodate and partition variability in these models, and capture covariances by fitting all parts of the model at once.
3. Frontiers in population forecasting are improved data fusion and making projections more iterative and automated.
4. Forecasting communities is challenging because of the large number of interactions, but not all such interactions will be equally important. While the possibility of unknown unknowns should be acknowledged, it should not prevent us from making projections.

5. It can be useful to identify the important scales of variability in a system before adding determinism to a model, and then to explore scales in order of importance

## 7.4 HANDS-ON ACTIVITIES

The code for the Florida scrub mint case study by Evans et al. (2010) is publicly available on Ecological Archives (http://esapubs.org/archive/mono/M080/022/) as well as on Github (https://github.com/mekevans/Dicerandra).

# 8

## Latent Variables and State-Space Models

*SYNOPSIS: In trying to understand and predict the world around us, a ubiquitous challenge is the inability to directly measure the variables we are interested in without uncertainty. In this chapter we will focus on this general class of problems where the variable of interest is latent, which indicates that it is unobserved or estimated with uncertainty. Particular attention is paid to latent variables that vary across time and space, as these are central to many forecasting problems. To address this problem I introduce a general statistical framework, known as the state-space model, for estimating latent variables in space and time subject to both observation and process model errors. This framework serves as the foundation for much of our discussion of data fusion (chapter 9) and data assimilation (chapters 13 and 14).*

### 8.1 LATENT VARIABLES

Latent variables are ubiquitous in ecological forecasting problems. This section focuses on four common examples of latent variable models: both random and systematic observation errors, proxy data, missing data, and unobserved variables.

We first encountered latent variable problems in terms of *observation error* (section 6.3). Observation errors introduce uncertainty in the state variables of a system that leads to us not knowing the true value of such variables. In the general case, we often think of the observed data, $y$, being related to the true, latent state of a system, $x$, according to some *observation model*, $y \sim g(x|\phi)$, where $\phi$ are the parameters of the model $g$. Observation models can have both random and systematic components. *Random observation errors*, $\varepsilon$, are typically modeled as a probability distribution with mean zero, for example, $y = x + \varepsilon$, where $\varepsilon \sim g(0|\phi)$. Furthermore, while random errors may be independent in space or time, in practice they are often autocorrelated (section 6.6). In contrast to random observation errors, *systematic errors* imply some *bias* in the measurement process. For the simple case of a constant additive bias, the combination of random and systematic errors might be captured by a relatively simple observation model, such as

$$y \sim g(x|\phi) = N(x - b, \sigma^2)$$

where $b$ is the bias and random errors are assumed to be Gaussian. In many cases systematic errors are not known a priori, but are estimated with uncertainty, and thus data cannot simply be "bias-corrected." Furthermore, systematic errors are not always constant, but can vary in space and time, sometimes driven by external variables we understand, while at other times responding to variables that are unmeasured or apparently random (such as sensor drift). For example, systematic errors associated with some known covariate $z$ could be modeled parametrically as a function of $z$, $f(z)$. By contrast, sensor drift might be approximated as a random walk process. The combination of these, plus random observation errors $\sigma^2$, might lead to an observation model

$$y_t \sim N(x_t - b_t - f(z_t|\phi), \sigma^2)$$
$$b_t \sim N(b_{t-1}, \tau^2)$$

(8.1)

where $\tau^2$ controls the rate of drift in the bias through time. Such a bias might be both estimated and accounted for using the state-space framework discussed in the next section. As discussed in chapter 6, in a model like equation 8.1 the variable we're interested in, $x$, would be unidentifiable using data on $y$ alone because $x$, $b$, and $f(z)$ would trade off and end up very tightly correlated. Constraining the model requires some combination of either strong prior information on $b$ and $f(z)$, additional data that specifically informs the systematic errors (for example, periodic calibration curves), or the fusion of multiple types of observations that helps identify the systematic inconsistencies in each of them (see chapter 9). Fortunately, data producers are becoming more aware of the importance of reporting both random and systematic uncertainties.

Another common example of latent variables is when the quantity we can measure is a *proxy* for the variable we are interested in, such as the earlier example where we used TDR as a proxy for soil moisture (section 6.3). In that example the observation model, $g$, was an empirical calibration curve. More generally, observation models can, like any other model, vary from simple statistical models to complex computer codes, such as the radiative transfer models used in remote sensing (Jacquemoud et al. 1995). However, even complex observation models can be dealt with in the same observation error framework as the calibration example. Furthermore, sometimes a single observation may be a proxy for multiple things, such as how $O^{18}$ in water is variously a proxy for temperature, evaporation, and atmospheric circulation (Wright et al. 1993); tree rings are used as a proxy for temperature, precipitation, or drought severity (Speer 2010); and animal telemetry may be a proxy for both metabolic activity and habitat preference. It is also true that we sometimes have multiple proxies for the same quantity, such as how forest inventory data, lidar, and radar have all been used as proxies for forest biomass, or how tree rings and satellite Normalized Difference Vegetation Index (NDVI) have both been used as proxies for net primary productivity (NPP). Indeed, except to the rare individual who is deeply interested in the absorption and reflectance of specific bands of electromagnetic radiation, all of remote sensing is based on generating and studying indirect proxies. Furthermore, some of the quantities we are most interested in, such as gross primary productivity (GPP), have nothing approaching a true direct measurement, but are estimated from often wildly different proxies ranging from flux tower observations, leaf-level gas exchange, MODIS reflectance, carbonyl sulfide (COS) uptake, and chlorophyll florescence. We can extend the framework we have already been discussing for observation error (chapter 6) to deal with these cases by explicitly including multiple data models (one for each data type) that all provide information about the same latent variable (chapter 9). We can similarly construct models where a single proxy observation provides information about multiple latent variables, though this is much more challenging due to issues of identifiability and strong covariances among latent variables. This type of problem will work best when such a proxy is combined with other observations (chapter 9).

In addition to observation error, we also encountered latent variables with *missing data models* (section 6.4). In this case a subset of the observations may be missing altogether, but can be estimated with uncertainty. In addition to the missing covariates problem discussed previously, missing data are a particularly common problem when looking across time and space—that is, that measurements are not made at all times and all places. As we shall see later, the state-space framework accounts for such missingness in space and time automatically.

Finally, not only can latent variable models be used to estimate variables that are observed with error, indirectly, or with missing data, but they can also be used to estimate *variables that are never observed at all*, but are merely inferred from the process model. One example of this was the latent seedbank in the Florida scrub mint case study (chapter 7). Resource allocation decisions by organisms represent another common ecological example of such latent variables. We frequently get to observe the outcome of such decisions, such as growth, fecundity, and allometry, but in most cases we have to indirectly infer the process (LaDeau and Clark 2006). Such models will be most successful when there are multiple data constraints (that is, the unobserved process has to be compatible with multiple independent measures of different aspects of the response) and when there is stronger a priori confidence in model structure (that is, mechanistic models rather than phenomenological ones).

## 8.2 STATE SPACE

While latent variable estimation problems are common and important unto themselves, for the rest of this chapter I want to narrow the focus to the common problem of latent variables connected in time or space. Such problems are in many ways the bread-and-butter of forecasting, since *the goal is typically to use a set of observations to predict key variables further in time or farther in space*, subject to the simultaneous problems of missing data (sparse in time or space), observation error, process error, and uncertainties in initial conditions, drivers, and parameters. Further examples and approaches for dealing with multiple proxies or multiple response variables simultaneously will be deferred to chapter 9. Here I want to introduce a useful framework for dealing with this class of problems called the *state-space model*, the name for which comes from its focus on estimating the latent state variables in the system. In the case of time-series analysis such models are also referred to as *Hidden Markov models*, with the "hidden" referring to the latent variable nature of the problem, and "Markov" referring to the fact that we will write such models recursively, with the next state in time a function of the current state. This is the same Markov property that we defined and used in Bayesian MCMC algorithms (section 5.6). The special case where the next state is a linear function of the current is also referred to as *dynamic linear model*. The state-space framework contrasts with classical time-series analysis, which involves a large covariance matrix between all observations simultaneously (section 6.6). The extension of the state-space framework to spatial modeling or space-time models is often called a *Markov random field* (MRF). Because of their capacity to flexibly capture and partition a wide range of uncertainties and address the complexities of real data (chapter 6), I rely heavily on state-space models as the basis of forecasting in space and time anytime the process model can be expressed as a dynamic model (that is, whenever the future state of the system depends upon its current state). Most ecological forecasting problems are likely to fall into this category.

## 8.3 HIDDEN MARKOV TIME-SERIES MODEL

To understand the Hidden Markov model, consider the example of a time-series analysis of seasonal influenza, which we will discuss in more detail in chapter 15. Specifically, we'll look at the flu index produced by the Google Flu Trends project (www.google.org/flutrends). Let's call $y_t$ our *observation* of the flu index at time $t$. Similarly, let's call the *true* latent state $x_t$, which in this case is the number of flu infections in the population. As earlier, $y_t$ is connected to $x_t$ through our observation error model, $y_t \sim g(x_t|\phi)$, with parameters $\phi$. So far this is just our standard observation error data model. Next, we'll model the evolution of $x$ through time using the dynamic process model $x_{t+1} = f(x_t|\theta) + \varepsilon_t$, where $\varepsilon_t$ is the additive process error and $\theta$ are the model parameters (figure 8.1). This type of recursive process model will look very familiar to anyone who's done any population or ecosystem modeling, and thus is broadly applicable to many ecological models and forecasting problems. Combining these two parts gives us the general state-space framework:

$$y_t \sim g(x_t|\phi)$$
$$x_{t+1} = f(x_t|\theta) + \varepsilon_t \tag{8.2}$$

When dealing with multiple time-series simultaneously (also known as repeated measures), it is common for $\varepsilon_{i,t}$ to have some hierarchical structure, such as having shared year effects among individuals, or time-persistent effects on individuals, plots, and so on. Similarly, it is not uncommon for $\theta$ to be modeled hierarchically to allow the model parameters to vary with time, individuals, plots, and so on.

   To make this a bit more concrete, let's start with the simplest possible dynamical process model, a random walk $x_{t+1} = x_t + \varepsilon_t$, and assume that both our process error and observation error are Normally distributed.

$$y_t \sim N(x_t, \tau_{obs}) \qquad \text{Data model}$$
$$x_{t+1} \sim N(x_t, \tau_{add}) \qquad \text{Process model}$$
$$1/\tau_{obs} \sim Gamma(a_{obs}, r_{obs}) \qquad \text{Observation prior}$$
$$1/\tau_{add} \sim Gamma(a_{add}, r_{add}) \qquad \text{Process prior}$$
$$x_0 \sim N(x_{IC}, \tau_{IC}) \qquad \text{Initial condition prior}$$

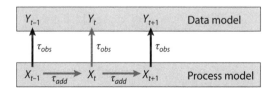

FIGURE 8.1. State-space model describing the evolution of the latent state variable, $X$, conditional on the observations, $Y$. In this random-walk example the only components are observation error, $\tau_{obs}$, and process error, $\tau_{add}$. The gray arrows indicate the connections relevant for estimating the posterior distribution for $X_t$.

Note that in addition to priors on the observation and process errors, there is also a prior on $x_0$, the initial condition.

To understand the power of the Hidden Markov model, consider the probability distribution for the state variable, $x_t$, conditional on the values of the model parameters

$$x_t| \ldots \propto \underbrace{N(x_t|x_{t-1},\tau_{add})}_{previous\ time}\underbrace{N(x_{t+1}|x_t,\tau_{add})}_{next\ time}\underbrace{N(y_t|x_t,\tau_{obs})}_{observation}$$

Here we see that our estimate of any $x_t$ depends upon the $x$'s that come *both* before and after, as well as upon the observation, $y_t$. Because this posterior is just a product of Normal distributions, the relative strengths of the data model versus the process model in constraining $x_t$ ends up being a precision weighted average, just as we saw the weighting between the data and prior when calculating a simple mean (see equation 5.7). In other words, when there is a lot of observation error, the model gets more weight, while when observation error is small, the data gets more weight. Similarly, when the deterministic core of the model explains a larger fraction of the observed variability, which leads to the process error being smaller, the weight of the model increases. The other thing that is noteworthy about this model is that the observed data points, $y$, are treated as *conditionally independent* given the latent variable $x$. This means that an observation does not depend upon any other observation, just the observation error model, while the dynamics of the system depends upon the process model linking the $x$'s, which is specified independent from observation error. This separability of observation and process error is part of what gives the state-space framework its strength and flexibility.

The flexibility of the Hidden Markov model can be seen by considering what happens when there are missing data or irregularly spaced data, a situation that can be challenging for traditional time-series models. If $y_t$, the observation at time $t$, is missing, then the posterior for $x_t$ becomes

$$x_t| \ldots \propto \underbrace{N(x_t|x_{t-1},\tau_{add})}_{previous\ time}\underbrace{N(x_{t+1}|x_t,\tau_{add})}_{next\ time},$$

This shows that we can estimate missing values based on the process model, but that the posterior will be of lower precision (recall that the precision of the posterior is the sum of the component precisions). In the case of multiple missing observations in a row, since the value that came before (or after) is also missing, and thus itself has higher uncertainty, the uncertainty compounds. Over larger gaps this causes the confidence interval to take on a balloon-like bulge, with uncertainty increasing the further you get from an observation (figure 8.2). This is very different from the pattern you get if you simply interpolate across data gaps, which treats missing data as if they are known perfectly without uncertainty. It also is very different than any sort of gap-filling algorithm based on covariates or average values, which may produce an uncertainty estimate, but one where the uncertainty right after the last measurement would be just the same as the uncertainty in the middle of a large data gap. Intuitively, that behavior is unappealing—we should be more uncertain the larger a gap is and the closer we are to the center of a gap. The Hidden Markov model demonstrates this intuitive behavior. State-space models can also handle irregularly spaced data automatically by simply treating the time between observations as missing data.

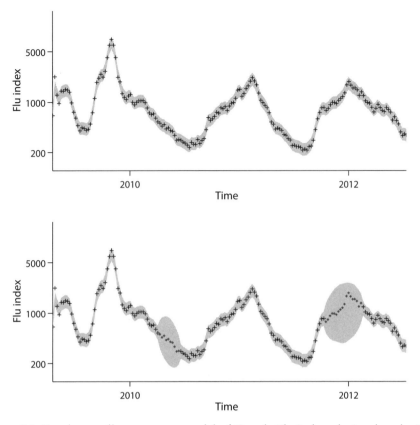

FIGURE 8.2. Random-walk state-space model of Google Flu Index, depicted as the 95% confidence interval (shaded area). Bottom panel includes two imposed data gaps of 10 and 18 weeks. Observations are depicted by black crosses, while missing data are indicated by diamonds.

A similar phenomenon occurs at the end of the time series. At the time of the last observation the posterior is

$$x_t| \ldots \propto \underbrace{N(x_t|x_{t-1}, \tau_{add})}_{\text{previous time}} \underbrace{N(y_t|x_t, \tau_{obs})}_{\text{observation}},$$

which likewise has a lower precision because there is no future observation to constrain the estimate. Similarly, the posterior for a forecast beyond the last observation is just

$$x_t| \ldots \propto \underbrace{N(x_t|x_{t-1}, \tau_{add})}_{\text{previous time}},$$

which, like the large gaps, will increase in uncertainty the further into the future the forecast is made (figure 8.3).

It should be noted that there is no requirement in the state-space framework that either the data model or process model be Gaussian. For example, the state-space framework has been applied to mark-recapture data with a Bernoulli observation

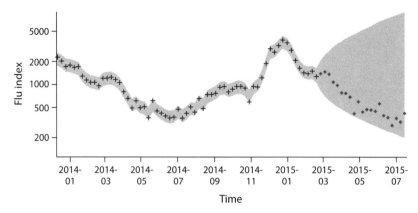

FIGURE 8.3. Forecast of the Google flu trend data compared to observations. Gray diamonds (starting March 2015) were not assimilated into the forecast.

error (probably of capture) and a Bernoulli process error (probability of survival) (Whitlock and McAllister 2009). Poisson or Negative binomial observation errors for count data, gamma or log-Normal models for zero-bound data, and multinomial observation errors for categorical data are all also straightforward.

In most cases the process model in the state-space framework is more complex than a simple random walk. Model complexity can range from the purely statistical (for example, linear models of the current state plus other covariates), to semi-mechanistic nonlinear functions (for example, population growth; see chapters 7, 10, and 15), to detailed mechanistic simulation models (for example, ecosystem models; see chapter 12). The only restrictions on the process model are that it be a dynamic model (that is, at least in part a function of the current state of the system) and that a process error be included in the overall framework (equation 8.2). That said, the random walk remains a useful "null" model to compare to more complex process models.

Since the presence of observation and process errors makes the state-space model so flexible at following observations, model diagnosis and model comparison require a bit more effort than just comparing predictions to observations. The size of the estimated process error is a key indication of how much of the observed variability was actually explained by the process model versus just apportioned to model error. It can also be useful to calculate, for every time step, the expected value of the process model (in the absence of process error) at the next time step, $E[f(x_t)]$. The difference between this and the actual posterior estimate of $x_t+_1$, expressed either as a residual or as a posterior probability, is a useful diagnostic when helping to identify where the process model may be failing. Chapter 16 will present approaches to model validation, benchmarking, and the identification of model structural errors in greater detail.

## 8.4 BEYOND TIME

This section extends the state-space model beyond just time, to consider space, space-time, and networks. Specifically, I present the key concepts of raster gridded spatial models (Markov random field, or MRF); vector polygon spatial models (conditional

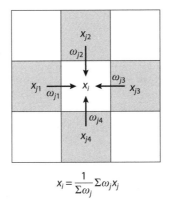

$$x_i = \frac{1}{\sum \omega_j} \sum \omega_j x_j$$

FIGURE 8.4. Process model for spatial random walk with a 4-cell neighborhood.

autoregressive, or CAR); and Bayesian networks, but defer the reader to other texts for extensive treatments of the details involved in these models (Banerjee et al. 2003; Cressie and Wikle 2011; Korb and Nicholson 2011).

To start, the simple 1-D time-series version of the state-space model can easily be extended to two (or more) dimensions. Doing so primarily requires that the process model be dynamic in space (that is, that it can be written as a function of the state of the system in adjacent locations), the same way that the time-series model was dynamic in time. Such a model may represent a genuine spatial process (for example, dispersal, spread, flow) or may just represent the hypothesis that the underlying spatial process is smooth. Most MRF models are run on a grid with some neighborhood adjacency defined (for example, 4 nearest neighbors or 8 nearest neighbors). By comparison to the Hidden Markov time-series model, the MRF tends to have a much higher fraction of missing data, which means that projections typically involve more uncertainty and the mean estimate is often smoothed more.

As with the Hidden Markov model, the MRF is easiest to explain with a simple example. The spatial equivalent to the random walk model would be to say that $x_i$, the state variable at location $i$, is a weighted average of the values in the neighboring locations, plus some process error (figure 8.4).

Writing the full model out gives

$$y_i \sim N(x_i, \tau_{obs}) \qquad \text{Data model}$$

$$x_i \sim N\left(\frac{1}{\sum \omega_j} \sum \omega_j x_j, \tau_{add}\right) \qquad \text{Process model}$$

$$1/\tau_{obs} \sim Gamma(a_{obs}, r_{obs}) \qquad \text{Observation prior}$$

$$1/\tau_{add} \sim Gamma(a_{add}, r_{add}) \qquad \text{Process prior}$$

$$x_{edge} \sim N(x_{BC}, \tau_{BC}) \qquad \text{Boundary condition prior}$$

Here you see that except for the different process model, the overall model is virtually identical. The only other notable difference is that instead of specifying a prior on the initial condition, *in a spatial model we instead specify a prior on the boundary condition*—the state of the system just outside the region we are interested in. Depending on the type of boundary condition imposed in the process model, the boundary condition prior may not be necessary (for example, reflecting or wrapping boundaries).

It should be noted that a spatial model need not be on a raster grid, but the same concept can be applied to any spatial network or vector map that can be represented as a graph (that is, as a collection of nodes that are connected by a set of edges). This class of models is closely related to the conditional autoregressive (CAR) model, which is frequently used not as a process model, but as a prior for spatial random effects. The overall structure of the model stays the same in the case of *networks or vector maps*; what changes is the weights, $\omega_i$, that are assigned to each connection. For example, in the original raster map the weights tend to be equal for all neighbors. By contrast, with a vector map the weights may vary as a function of the size of each polygon, the length of their border, the distance between centroids, or any other ecologically meaningful metric of how one polygon influences another. For spatial networks these may involve the distance between patches, the size of patches, river networks, and so on. The underlying core concept is that if we have any ecologically meaningful metric to define the connections (for example, adjacency) between our latent $x$'s, and assign weights to those connections, then we can utilize the flexibility of the state-space framework to estimate the latent $x$'s, separate observation and process error, and deal with missing data. Finally, *the use of state-space models to describe networks need not be limited to spatial networks*. The same framework works equally well for food webs, phylogenetic relationships, kin or social relationships, biogeochemical networks, and so on. Indeed, there is an extensive literature on Bayesian network analysis that could find substantial use in ecological forecasting (Korb and Nicholson 2011).

## 8.5 KEY CONCEPTS

1. Common examples of latent variable models are both random and systematic observation errors, proxy data, missing data, and unobserved variables.
2. State-space models treat the state of the system as a latent variable to be estimated and in doing so explicitly separate observation errors from process errors.
3. Because of their capacity to flexibly capture and partition a wide range of uncertainties and address the complexities of real data, I generally recommend state-space models as the basis of forecasting in space and time any time the process model can be expressed as a dynamic model (that is, whenever the future state of the system depends upon its current state).
4. The observed data points, $y$, are treated as conditionally independent given the latent variable $x$, which means that an observation does not depend upon any other observation, just the observation error model, while the dynamics of the system depends upon the process model linking the $x$'s, which is specified independent from observation error. This separability of observation and process error is part of what gives the state-space framework its strength and flexibility.
5. Missing data gaps and irregularly spaced data are handled automatically, with uncertainties increasing with distance to the nearest observation.
6. State-space models require a prior on the initial condition (time) or boundary condition (space).

7. State-space models can be extended beyond time and gridded space to also consider vector maps, spatial networks, and functional networks (food webs, phylogenetic relationships, social relationships, and so on).

## 8.6 HANDS-ON ACTIVITIES

https://github.com/EcoForecast/EF_Activities/blob/master/Exercise_06_StateSpace .Rmd

- Random walk time-series model

# 9

## Fusing Data Sources

*Synopsis: Making forecasts commonly involves combining information from multiple sources. Bayesian techniques provide new ways to do this rigorously, but balancing the information provided by different data sources remains among the most debated topics in ecological model-data fusion. Here I discuss why and how data sources can be combined, lay out successes and failures, and highlight the areas in need of further research.*

AT ITS HEART, this chapter is about *synthesis*. Synthesis is an essential component of the scientific process and is particularly important for forecasting, where we aspire to make the best predictions possible. However, it is often the case that no single data set provides a complete picture of the system we are interested in. To make the most accurate and informed forecasts, we typically want to use all available data, rather than relying on just one line of evidence. This frequently requires that we bring together information from different sources.

The use of multiple data sources can be particularly powerful in addressing the issue of *identifiability* raised in chapter 6. Consider, as an example, the carbon balance of an ecosystem (figure 9.1), which can be summarized according to the simple mass balance that net ecosystem production (*NEP*) is gross primary productivity (*GPP*) minus autotrophic ($R_a$) and heterotrophic ($R_h$) respiration:

$$NEP = GPP - R_a - R_h$$

Eddy-covariance is commonly used to estimate *NEP*, but it should be clear that the terms on the right hand side cannot be uniquely identified from such data (Williams et al. 2009). Commonly, nighttime measurements (where *GPP* = 0) are used to partition *GPP* from ecosystem respiration ($R_e = R_a + R_h$) (Reichstein et al. 2005), but doing so requires the introduction of a process model for $R_e$, a data model, and the assumption that the relationship between environmental covariates and fluxes is the same for daytime and nighttime fluxes, which is not always valid (Phillips et al. 2011; Munger et al. 2016).

However, it should also be clear that if *NEP* is constrained by data, there will be strong statistical covariances among the three component fluxes, even if they are not individually identifiable. What that means is that independent information on the component fluxes, such as biometric measurements of net primary productivity ($NPP = GPP - R_a$), will serve to constrain the other term with which it covaries. To

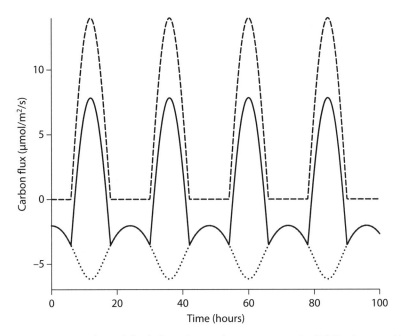

FIGURE 9.1. Conceptual model of the relation between *NEP* (solid line), *GPP* (dashed line), and *RE* (dotted line).

revisit the concepts from chapter 5, this is just saying that if we have a complex covariance structure (for example, figure 5.9, "banana"), and one variable can be estimated precisely, then the uncertainty about the other variable is the much narrower conditional distribution, rather than the broad marginal distribution. That said, information from different data sources can at times be redundant (Keenan et al. 2013), and when faced with limited resources for new measurements, or the synthesis of old measurements, it is important to prioritize effort to the data that provide the greatest constraint.

While the importance of synthesis is clear, putting this into practice is not as straightforward. Fusing data involves more than just concatenating data files together. Naïve interpolation or extrapolation from one data type to another, without a rigorous accounting of uncertainty, can lead to biased and overconfident results. Traditional statistical models are of limited use when an analysis involves multiple processes that may be observed at different scales, locations, and time points, and that have different sources of uncertainty. You won't find predefined functions in your favorite statistics software that can account for the complexity of such analyses. Unfortunately, independent analyses of different data sets are frequently cobbled together, and in the process throw out the uncertainties. When uncertainties are ignored or approximated, usually the first thing to go is the covariance matrix, which is unfortunate because, as we saw in the *NEP* example, it is the covariance that provides much of the statistical and inferential power. As data availability increases, the need for best practices in combining data sources into a single analysis or forecast becomes more and more important. This chapter focuses on the tools, techniques, and pitfalls associated with bringing multiple data sources together.

### ▦ 9.1 META-ANALYSIS

*Meta-analysis* is the form of quantitative synthesis most familiar to ecologists. The goal of meta-analysis is to *combine information, usually in the form of summary statistics, from independent studies.* The primary advantage of meta-analysis is that, by combining like information across studies, we are able to increase statistical power. Indeed, there are classic examples from the literature where the joint inference across many small studies, each of which was nonsignificant or demonstrated a weak effect, was able to detect a strong significant effect (Lau et al. 1992, 1995). Meta-analysis is also significantly more robust than other forms of literature synthesis, such as a qualitative review or a "vote-counting" approach (that is, counting the number of papers demonstrating an effect), because different studies will have very different sample sizes and uncertainties and thus should not be counted equally (Koricheva et al. 2013). Figure 9.2 illustrates an ecological example of a cumulative meta-analysis, which shows how the combined evidence up to that point in time changes as more studies accumulate.

Meta-analysis often focuses on data of a single type or response variable, which is quantified through a metric of *effect size*, that indicates the strength of the relationship observed within each individual study. Common measures of effect size include the absolute or percentage difference between the means of two groups, correlation coefficients, and regression slopes. For any effect size statistic, error estimates and sample sizes are also critical, though unfortunately these are less frequently reported in the literature.

While less common, meta-analysis can also be used to directly estimate priors for parameters in a larger model (LeBauer et al. 2013). Here the effect size statistic is the parameter itself, such as a stoichiometric ratio (for example, C:N). Compared to parameterizing a model from a single study or site, a meta-analysis provides both greater constraint and the ability to account for the real ecological variability among multiple studies. Constraining parameters through meta-analysis also allows uncertainty analyses (chapter 11) to better identify parameters that need additional measurement. By requiring parameters to be consistent with literature observations, it also reduces the risk of calibration (chapter 5) producing biologically unrealistic parameters.

When producing a high-quality meta-analysis, it can be challenging to ensure that the analysis is rigorous and the sample of data used is as complete as possible (that is, that studies were not missed). Most notable of these challenges is the *reporting bias* problem, *that neutral or negative results are less likely to be published than positive results.* While most familiar in the meta-analysis context, reporting bias is a risk for all types of synthesis. Techniques exist to estimate the size of this bias and provide some degree of correction—for example, by exploiting the relationship between sample size and effect size (for example, funnel graph). However, such techniques need to be used with caution to ensure that any specific problem conforms to the missing at random condition for missing data problems (chapter 6). A bit more tractable is the problem of partial missing data, as publications that include means or other effect sizes are frequently missing information about the uncertainty in such data or the sample size. Using a combination of conservative assumptions (for example, minimal sample sizes) and missing data models, it is often possible to include studies that are missing this data (LeBauer et al. 2013), but because of the need to

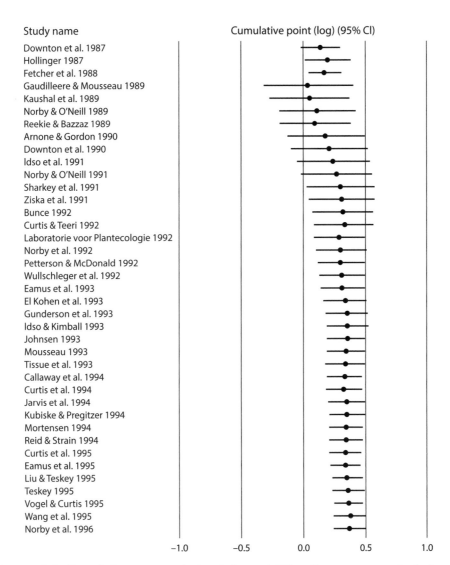

FIGURE 9.2. Cumulative meta-analysis of elevated $CO_2$ effects on net assimilation in woody plants. Data from Curtis and Wang (1998); figure from *Handbook of Meta-analysis in Ecology and Evolution*, edited by Julia Koricheva, Jessica Gurevitch, and Kerrie Mengersen. Copyright © 2013 by Princeton University Press. Reprinted by permission.

integrate over uncertainty, such data points obviously provide far less information than complete studies. Anyone thinking of conducting a meta-analytical study is encouraged to study the tools, techniques, and challenges in greater detail (Koricheva et al. 2013).

There are a number of statistical modeling approaches available for meta-analysis. What distinguishes a meta-analytical model from other models is that the observations are typically summary statistics, and thus there is a need for *different studies to have different weights based on their sample sizes and variability*. Here I describe a simple Bayesian hierarchical model that is used to differentiate within-study versus

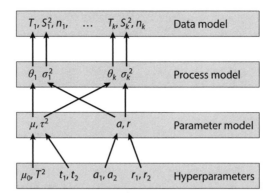

FIGURE 9.3. Hierarchical Bayes meta-analysis model. In a meta-analysis the data are not raw observations but are summary statistics from each of $k$ publications: sample mean ($T$), sample standard deviation ($S$), and sample size ($n$).

across-study variability and to account for the *sampling error* that arises from working with summary statistics. Let's assume that the $i^{th}$ study has a true mean, $\theta_i$, and a true within-study variance, $\sigma_i^2$, but these quantities are latent—we don't get to observe them directly. Instead we get to observe $T_i$, the sample mean, and $S_i$, the sample standard deviation (not standard error) (figure 9.3). If we assume the samples are Normally distributed, then

$$T_i \sim N(\theta_i, \sigma_i^2/n_i)$$

$$\frac{1}{S_i^2} \sim Gamma\left(\frac{n_i}{2}, \frac{n_i \sigma_i^2}{2}\right) \tag{9.1}$$

where $n_i$ is the sample size. It is important to note that the $n_i$ shows up in the denominator of the first equation, since $T_i$ is a mean of $n_i$ observations. The second equation arises from equation 5.8, where we showed that the precision is distributed according to a Gamma distribution parameterized as half the sample size and sum of squares. At the next level up in the hierarchical model, the true means are assumed to be Normally distributed with an across-study mean, $\mu$, and an across-study variance, $\tau^2$

$$\theta_i \sim N(\mu, \tau^2)$$

When estimating the across-study mean, not all individual studies contribute equally. Recall from chapter 6 that $\mu$ will be a precision weighted mean of the $\theta_i$, and thus from equation 9.1 we can see that the weight provided from an individual study will increase when either the sample size is large or the standard deviation is small. The study-specific within-study precision follows a Gamma distribution

$$\frac{1}{\sigma_i^2} \sim Gamma(a, r)$$

To assume a common within-study variance, instead of a study-specific one, the $i$ subscript simply needs to be dropped. Alternatively, the within-study variance can be made to have a hierarchical constraint by specifying priors on $a$ and $r$. Given the frequency at which studies fail to report sample standard deviations, $S_i$, or other error statistics, the choice of either a common or a hierarchical within-study variance is recommended. These alternatives allow the missing data model on $S_i$ to borrow strength across studies, with a common parameter being the stronger (and stricter) assumption.

## 9.2 COMBINING DATA: PRACTICE, PITFALLS, AND OPPORTUNITIES

Hands-down, the most straightforward way to bring external information into an analysis is as a prior. That said, the conditional nature of Bayesian modeling makes it relatively straightforward to build models that contain multiple data models (likelihoods), and there are many cases where this makes more sense than performing the analysis sequentially. If you imagine likelihoods like Lego blocks, you can build up very complex statistical models, which synthesize a wide range of observations of different types and inform different parts of a process, by putting these simple building blocks together. Like Legos, when combining likelihoods, it's important for the components to have parts that fit together. For example, parts may be combined based on conditional probabilities: $P(A|B) \cdot P(B|C) \cdot P(C)$. Alternatively, parts may involve different data sets that share the same process models: $P(X_1|\theta) \cdot P(X_2|\theta) \cdot P(X_3|\theta) \cdot P(\theta)$. Occasionally, a single data set might inform multiple processes: $P(X|\theta_1) \cdot P(X|\theta_2) \cdot P(\theta_1) \cdot P(\theta_2)$. For example, the Florida scrub mint case study in chapter 7 had eleven different process models, five different types of data, and scores of priors that all fit together. Some parameters were unique to one submodel, while others, such as the shared YEAR effect, were common across all parts.

Within an MCMC, updating any particular parameter doesn't require knowing what the whole model is doing, just the parts that contain that one parameter, as updating is done conditional on the current state of everything else in the model. In other words, there are no new tools or theories required to build complex models synthesizing many data streams. This property (being able to build and sample complex models from simple parts), in combination with the updatable nature of Bayesian inference, provides an enormous opportunity for data synthesis. That said, there are definitely potential pitfalls for the unwary, and challenges that lay before the community on how best to address these pitfalls, which themselves represent new research opportunities. The rest of this section will largely focus on these pitfalls and the current best practices to address them.

The simplest case of data fusion would be when multiple independent data sources, $\vec{Y}_i$, inform the same process model, $f(x|\theta)$, each through its own data model, $g_i$

$$
\begin{aligned}
\mu &= f(x|\theta) \\
\vec{Y}_1 &\sim g_1(\mu|\phi_1) \\
\vec{Y}_2 &\sim g_2(\mu|\phi_2) \\
&\vdots \\
\vec{Y}_k &\sim g_k(\mu|\phi_k)
\end{aligned}
\tag{9.2}
$$

A simple example would be when there are different methods to measure the same quantity, each unbiased but with different standard deviations. As shown in chapter 5 (equation 5.7), if the process model were just the mean, then the posterior mean would be the precision-weighted average of the different data sets, and the posterior precision would be the sum of the precisions. This reminds us that even if there is a single "best" data set, the synthesis across multiple data sets will always be more precise.

As a concrete illustration, consider trying to use two data sets, $(X_1,Y_1)$ and $(X_2,Y_2)$, to constrain the same linear regression model. Let's further assume that these data sets don't have the same observation errors—one method is more precise but also more expensive, and thus has a smaller sample size, while the other is less precise but more common. Neither of our conventional regression approaches works well in this case; it would be inappropriate to combine the data into one large data set because of the unequal variance, but fitting the data separately would just give us two distinct regressions rather than a synthesis. Instead we can write down two likelihoods, one for each data set, that have the *same* process models but different data models (specifically, different variances):

$$Y_1 \sim N(\beta_0 + \beta_1 X_1, \sigma_1^2) \quad \text{Likelihood 1}$$

$$Y_2 \sim N(\beta_0 + \beta_1 X_2, \sigma_2^2) \quad \text{Likelihood 2}$$

$$1/\sigma_1^2 \sim Gamma(a_1, r_1) \quad \text{Prior error 1}$$

$$1/\sigma_2^2 \sim Gamma(a_2, r_2) \quad \text{Prior error 2}$$

$$\beta \sim N_2(\mu_0, V_0) \quad \text{Regression prior}$$

When we fit this model to data (figure 9.4), we see that the combined model produces a regression line that is between the independent fits and has lower uncertainty (tighter CI) than either fit alone.

Another form of synthesis occurs when multiple data sets are synthesized within a *multivariate model*. Here *multivariate* refers to a process model predicting multiple *output* quantities, which is in contrast with the types of models ecologists typically consider in multivariate statistics (for example, multiple regression, PCA and other ordinations, clustering, neural networks, support vector machines). Where this is likely to be encountered is with a process model that represents multiple interrelated mechanisms or structured models, such as age- or stage-structured population mod-

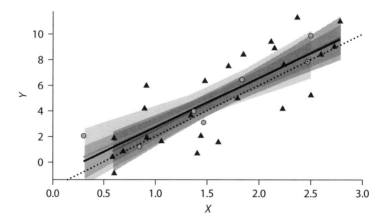

FIGURE 9.4. Fusing two regression models. Comparison of independent fits to independent data (gray lines: diamonds = common but noisy; circles = precise but expensive) and combined fit (black line).

els, dynamic models of ecological communities, or multiple-pool biogeochemical models. For example, a demographic analysis (chapter 7) might involve independent data sets related to current population size, growth and survival at different life history stages, reproduction, dispersal, rates of disturbance, habitat quality, and so on.

In generic terms, consider the case of $k$ data sets informing $n$ processes. Formally, there's no requirement that $k$ be greater or less than $n$; some processes may be informed by multiple data sets, while others might not be informed by any data at all. As with state-space models (chapter 8), because of the covariances in the system, it is possible to produce estimates for processes that are not observed directly (for example, Florida scrub mint seedbank, chapter 7); however, the uncertainty about a process increases the further it is removed from data. It is nonetheless possible to fit models with many latent variables, though problems of nonidentifiability are common. Also common is the opposite problem, where a tight constraint on one process, or through one data set, constrains another variable in an unrealistic manner. In both cases, it is critical to remember that *it is the structure present in the process model that determines the covariances among different processes.* This is directly analogous to how an autocorrelation function provides structure in a spatial or temporal autocorrelation model (chapter 6), except in this case we're relying on the hypotheses embedded in our process model to provide the covariance. However, since no process model is perfect, it remains important to include process error in the model.

A common problem when synthesizing multiple data sets is having data types of very different sample sizes. This often arises when some data sets come from high-volume automated sensors, such as data loggers making measurements at high temporal resolution or remote sensing data that provides large volumes of data across space, while other data come from low-volume "muddy boots" field measurements. Based simply on sample sizes, these automated data can overwhelm the often hard-won field data. Given that the field data usually contains information that is judged to be critical to capturing the process, this frequently drives researchers to adopt *ad hoc measures to balance the data.* Examples include ascribing arbitrary weights to the likelihoods for different data sets (for example, all data sets are reweighted to contribute equally to the fitting) (Medvigy et al. 2009; Richardson et al. 2010; Keenan et al. 2013), or resampling or averaging high-frequency data (for example, monthly means) so that they don't overwhelm the analysis. While such approaches frequently produce results that researchers intuitively feel are more sensible and balanced than throwing in all the raw data equally, they are intellectually unsatisfying. Adding arbitrary weights to likelihoods destroys our ability to interpret them as probabilities. Adding weights or thinning data has a large and subjective impact on the uncertainties associated with parameter estimates, rendering such estimates falsely over- or underconfident. Resampling or averaging large data sets throws out information and the weight given to different data sets likewise depends upon subjective choices about the degree of data reduction. For example, the decision on whether to work with raw 30-minute eddy-covariance data ($n = 17,520$/year), or to aggregate the data up to daily ($n = 365$), monthly ($n = 12$), or annual ($n = 1$) averages has a huge impact on the sample size, and thus the magnitude of the variance and the weight given to that data in the analysis.

While there is no magic bullet for combining data sources of different sizes, the current best practice is to treat the process and observation uncertainties associated with each data source appropriately (errors in variables, missing data, autocorrelation, heteroskedasticity, bias, and so on), as well as the uncertainties in the model (random and systematic process error), as failing to do so greatly exacerbates these problems. In particular, the information content of automated data may be greatly overestimated if observations are treated as independent. This doesn't just apply to temporal autocorrelation, but also to spatial dependence and other forms of autocorrelation, observation error, process error, and all other forms of uncertainty discussed heretofore (chapters 2, 6, and 8). If this information is being used to inform other locations, there will be spatial autocorrelation and random effects to address, which themselves may be scale dependent. These best practices don't just apply to the problem of fusing data; fusing data just highlights a problem that is present at all times because it creates a greater potential for internal inconsistencies either within the data (that is, systematic observation errors) or in the model (that is, systematic model errors).

As a simple example, imagine a temperature sensor set to record data every five minutes, but that only measures one location. When building the data model we would definitely want to account for the autocorrelation in the data, because the 288 observations/day we're getting are clearly not going to be independent. Next, we'd want to explicitly account for the calibration of the sensor. This is an errors-in-variables problem (section 6.3) because the sensor is actually measuring millivolts, not degrees, and the number of observations used to develop the calibration curve will be vastly smaller than the sample size from the sensor. After that we might want to account for the fact that the calibration is not time-invariant but may drift over time, introducing a systematic bias that might be informed based on periodic recalibrations (section 8.1). Furthermore, the temperature sensor might have systematic errors—for example, when it's in full sun or when it's wet—so we might want to develop a model of systematic bias as a function of light and precipitation. Screening out such data is also an option, but needs to be done with considerable caution because such data would then be missing systematically rather than missing at random. Next, we need to account for the spatial sampling error. However, since we have only one sensor (lack of replication), we can't estimate the sampling error from data and thus this error enters as an additional, unknown parameter in the bias. Here we might rely on similar experiments conducted in similar environments, but with multiple temperature sensors, to construct an informed prior on this term, which otherwise will be hard to identify.

The problem with both systematic errors and sampling biases is that they don't average out over time. Imagine the temperature sensor has a simple 5% systematic bias and a 20% random error. The naïve approach would consider the systematic error negligible compared to the random error, but if temperature is logged every 5 minutes, then the random error in the annual mean temperature is $20\%/\sqrt{105192} = 0.06\%$, while the systematic error is still 5%.

All of this just describes the challenges for the data model. Depending on how data are being used there may need to be additional accommodations in the process model as well. If temperature is the input data to another model, the observation data model may be sufficient, but if we're trying to predict temperature (for example, air or soil temperature in a biophysical land surface model or body temperature in an ecophysiological model), then we would need to account for both random and systematic process errors in the model as well; otherwise, the large volume of temperature data may drive the overall model calibration. Without doing so the model might be calibrated to do a great job at capturing temperature, but this might actually lead it to perform worse for other output variables (for example, soil moisture). Any errors in model structure would force the model to introduce compensating errors in other areas (that is, the calibration could force the model to get the right answer for the wrong reason).

## 9.3 COMBINING DATA AND MODELS ACROSS SPACE AND TIME

In this section, the process for data synthesis is extended to the state-space framework to deal with time-series and spatial data. This extension is fairly straightforward, given the latent variable structure of the framework that separates observation and process error. In the simplest case (analogous to equation 9.2), multiple data sources may inform one latent process, $f$, in which case each data set is just assigned its own data model, $g_i$.

$$x_{t+1} = f(x_t|\theta)$$
$$Y_{1,t} \sim g_1(x_t|\phi_1)$$
$$Y_{2,t} \sim g_2(x_t|\phi_2) \tag{9.3}$$
$$\vdots$$
$$Y_{k,t} \sim g_k(x_t|\phi_k)$$

As noted throughout this book, the process model, $f$, could range from the simplest null model (for example, random walk in time, diffusion in space) up to sophisticated mechanistic simulation. One of the obvious strengths of the state-space approach is that it flexibly combines measurements with differences in missing data—for example, observations made at different frequencies. When multiple observations are made, the estimate of the latent state combines information from all observations as well as the forward and backward constraints of the process model. When no observations are made, the process model alone provides constraint, as before. If whole data sets are separated in time (or space)—for example, due to switches in instrumentation—then care must be taken to not mistake changes in the underlying process for changes in observation error, as a lack of overlap can make these differences nonidentifiable.

It is also possible to construct state-space models that have multiple latent variables, such as different ages or stages in a population model, different pools in a biogeochemical model, different locations in a spatiotemporal model, or different measurement units through time (that is, repeated measures) that are connected through a hierarchical model. As with the static case discussed previously (section 9.2), some state variables may have multiple observations, while others have none and are only constrained through the process model.

A common challenge in data synthesis is combining data sets across different scales in space or time (figure 9.5). Far too often, rather than building an explicit spatial or time series model, data are moved to a common scale by naïve *interpolation* (for example, linear or cubic interpolation, splines, lowess), which throws out the uncertainty associated with the interpolation itself. If the common scale is at a fine spatial or temporal resolution this can result in an artificial inflation in sample size, leading to results that are falsely overconfident, while moving to a coarse resolution throws out information. Furthermore, information available at a coarser scale frequently

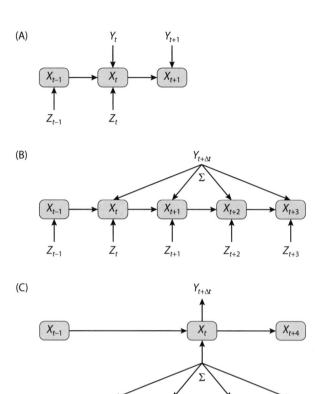

FIGURE 9.5. State-space fusion of multiple data sets, Y and Z, informing the same state, X. (A) observations at a common time step, but possibly different missingness. (B) Modeling at the resolution of the fine time-step data; multiple latent variables inform one observation Y. (C) Modeling at the resolution of the coarse time-step data; multiple Z observations inform one latent state.

represents an integral or average over space or time, meaning that naïve interpolation frequently causes a disconnect between averaged and instantaneous measurements, or integrals over different scales.

One way of addressing data on different scales is through a process model that explicitly represents these different scales. For example, a vegetation model might calculate carbon, water, and energy fluxes at a subdaily time scale, but calculate growth, mortality, reproduction and disturbance at a monthly or annual time scale (Medvigy et al. 2009). Using such a model as the process model in a state-space framework, it would then be very natural to utilize one set of measurements on a subdaily timescale (for example, eddy-covariance, soil respiration, sap flux, and so on), a different set of measurements at an annual scale (for example, tree rings, forest inventory), and a third set over successional timescales (for example, fossil pollen and charcoal). As noted earlier, care would need to be taken in constructing the data models for such an analysis, as the higher-frequency data will have a larger sample size, which, all else being equal, would tend to give it higher weight.

If one way to address the scaling problem is through the process model, the other way is to explicitly represent the time- or space-integration process as part of the data model. The most common way to do so is by numerically approximating the integration process by summing or averaging over the process model (figure 9.5). For example, in remote sensing it's not uncommon to have different platforms measuring comparable parts of the electromagnetic spectrum at different spatial and temporal resolutions because of inherent design trade-offs (for example, high frequency versus high resolution). For instance, Landsat and MODIS both measure multiple bands of visible and near-infrared radiation reflected off the earth's surface, which can be used to derive comparable vegetation indices (for example, NDVI, EVI) and estimates of ecologically interesting quantities, such as albedo, land cover class, and leaf area. However, Landsat measures the earth at ~30 m resolution and ~16-day frequency, while MODIS has a resolution of 250 m, 500 m, or 1000 m (resolution varies with spectral band) and a daily frequency. Both satellites frequently encounter clouds, so many derived MODIS data products represent a *time-average* over a 8- or 16-day window, while the equivalent Landsat estimate would be *instantaneous*. To combine these to estimate any particular variable (for example, NDVI) over a landscape, we might construct a process model that operates at the Landsat spatial resolution (30 m) and the MODIS temporal resolution (daily). The data model for Landsat data would thus have missing data 15 out of 16 days, in addition to missing data for clouds. MODIS NDVI is a 16-day average, 250 m product, so the data model for MODIS would need to average over both space and time. It should be clear that using a single MODIS observation results in a massive parameter identifiability problem—there are an infinite number of ways to get the correct mean observation given ~69 30 × 30 m pixels. Landsat alone would be slightly more manageable, but uncertainty would balloon between observations in time as we saw in chapter 8. However, each data set could borrow strength from the other, one providing temporal resolution and the other spatial. What is important to note, however, is that this approach isn't just combining the data sets arithmetically, assuming that one provides the correct temporal pattern and the other the correct spatial pattern, but rather is addressing the uncertainties in the underlying state variables. Furthermore, it should be clear that the choice of process model, which is what connects observations in space and time and thus allows us to borrow strength across data sets, will have a significant impact on the overall inference.

In addition to combining measurements across time, it is frequently the case that we need to combine measurements across space that are not all at the same resolution or even of the same geometry (for example, polygon verses raster). This is known as the *spatial misalignment problem* in spatial statistics and can be addressed in the same way as the previous example (Banerjee et al. 2003). For example, we might have environmental data on a raster grid, and need to compare this to census data that is associated with polygons. The process model would operate at a common raster spatial resolution and describe the relationship between the two variables, as well as the spatial autocorrelation in each process. The data models would then integrate from the common spatial resolution to the resolution of each data set.

An alternative approach to working at the finest spatial resolution would be to bring everything to the coarser resolution, but capture the fine spatial dynamics in a spatially implicit matter. This approach is computationally much less burdensome, but involves a loss of spatial information, and thus works best when fine spatial inference is not necessary for the question at hand or the forecast being produced. In the remote sensing example, we might work at the MODIS spatial resolution, but construct a process model that describes the statistical distribution of the underlying process (Stoy and Quaife 2015). As before, this distribution would be unidentifiable from a single MODIS pixel. The Landsat observations would then update the distribution, but the spatial locations would no longer be preserved.

Overall, there are abundant opportunities available for using state-space models to combine information across temporal scales, spatial resolutions, and processes, as well as opportunities for research into approaches to better accommodate the sample size differences inherent in working across scales and processes.

## 9.4 KEY CONCEPTS

1. It is often the case that no single data set provides a complete picture of the system we are interested in. Both inference and forecasting can be improved by bringing together information from different sources.
2. Multiple data sources can be particularly powerful in addressing the issue of identifiability.
3. Fusing data involves much more than concatenating files together or interpolating data to a common scale. Capturing uncertainties is critical to avoid bias and overconfidence, while covariances are critical to leverage complementary data types.
4. Data fusion frequently highlights systematic model errors.
5. Meta-analysis combines information, usually in the form of summary statistics, from independent studies.
6. Being able to build and sample complex models from numerous simple parts is key to data synthesis.
7. When combining likelihoods, avoid ad hoc and subjective choices, such as weighting likelihoods or thinning/averaging data. Instead, account for the complexities of the data (errors in variables, missing data, autocorrelation, heteroskedasticity, bias, observation error) and the model (random and systematic process errors).
8. Combining data often highlights systematic errors in models and data.
9. State-space models can allow us to combine spatial and temporal information that operate at different scales, even if these scales are misaligned.

## 9.5 HANDS-ON ACTIVITIES

https://github.com/EcoForecast/EF_Activities/blob/master/Exercise_08_TreeRings.Rmd

- Fusing times-series data: tree rings and forest inventory

# 10

## Case Study: Natural Resources

*Synopsis: Natural resource managers have been making "ecological forecasts" for decades. What have been the major successes and failures? Where are there opportunities for improvement?*

No area of ecological forecasting has a longer history than natural resources. Some of humankind's earliest innovations in time-keeping, astronomy, and mathematics were driven by the need to keep track of phenological cycles. In modern times empirical growth and yield curves have been in use for at least a century (Blackman 1919). Similarly, process-based plant growth and forestry models have been used to make projections since the 1960s (Bugmann 2001; El-Sharkawy 2011). Since our focus is on ecological forecasts, within the area of natural resources we shall limit our discussion to biotic resources, such as forestry, hunting, and fishing.

Natural resource forecasting is probably the most advanced in aquatic systems. This is somewhat ironic, because terrestrial plants and animals are generally easier to observe than fish and other aquatic organisms. However, it is precisely because most variables are latent that forecasting aquatic systems has required more sophistication. Similarly, it is likely because fish are harder to observe that such a large number of modern fisheries have collapsed, thus creating a strong financial and policy driver for fishery stock assessments and projections.

### 10.1 FISHERIES

I live in a small village on the Massachusetts coast south of Gloucester; thus, in many ways the story of American fisheries begins in my own backyard. Settled in 1623, Gloucester has been at the center of US commercial fishing in the north Atlantic for centuries, and that fishing has long centered on the Atlantic cod (*Gadus morhua*). Historical records tell of an inexhaustible bounty of cod in the region—a natural abundance so significant that Mark Kurlansky's popular books on cod (Kurlansky 2010) and Gloucester (Kurlansky 2009) refer to it as "the fish that changed the world." Indeed, Kurlansky argues that Basque cod fishermen actually discovered North America before Columbus, but like any fisherman with a favorite "secret spot," failed to disclose its location to others. As an example of how important cod was to the early New England economy, a 5-foot-long wooden "sacred cod" was hung in the Massachusetts State House in 1784 and is still there today.

Advances in fishing technology introduced in the post-war era led to large catch increases. However such levels of exploitation proved to be unsustainable, leading to

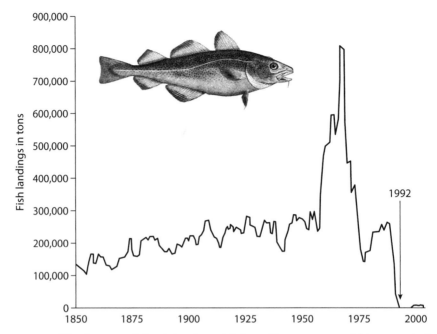

FIGURE 10.1. The history of the Atlantic cod fishery illustrates a rapid increase in landings in the post-war era of fishery industrialization. This led to a peak in the 1960s, a decline in the 1970s, and a collapse in 1992 that led to the closure of the Newfoundland stock (Millennium Ecosystem Assessment 2005).

a fishery collapse in the early 1990s (figure 10.1), and Atlantic cod is now listed as Vulnerable on the IUCN Red List. This collapse devastated the economies of fishing towns across New England and eastern Canada, sparking tensions between fishermen, regulatory agencies, and conservation organizations that continue to this day.

While the story of the Atlantic cod is perhaps the most famous example of fishery collapse, and an archetype for the Tragedy of the Commons (Hardin 1968), it is unfortunately not the only example. The Food and Agriculture Organization of the United Nations estimates that over 90% of marine fisheries are either fully fished or overfished (FAO 2014). As global population continues to rise, demand for fish is likewise only going to go up, and with it the need for accurate stock assessments and forecasts that can be used to set ecologically, economically, and culturally sustainable harvest levels. This goal is further complicated by concurrent changes in other environmental variables such as ocean temperature, pH, nutrient inputs, oxygen, and pollutants. These changes have both direct impacts on fish and indirect impacts through changes in marine NPP and marine community shifts in response to overfishing (for example, increases in jellyfish).

Most fishery modeling is based on population modeling principles, expanding on the basic themes discussed in chapters 2 and 7. Fisheries traditionally use age-structured models with slightly different functional forms than terrestrial ecologists, such as the Beverton-Holt model instead of the logistic model to describe density-dependence (Beverton and Holt 1957; Hilborn and Walters 1991). As with much of ecology, in fishery models there is a continual tension between those arguing for simple models versus those advocating for realistic biological detail (Kareiva et al. 2000; Mann and Plummer 2000).

While the models used by fisheries biologists are not radically different from those used in other ecological subdisciplines, what really sets fisheries apart is a much longer history of using Bayesian methods (Punt and Hilborn 1997). Indeed the first author of *The Ecological Detective* (Hilborn and Mangel 1997), a classic book on statistical modeling that introduced many ecologists (myself included) to maximum likelihood and Bayesian methods (chapter 5), is the same fisheries ecologist who wrote the standard textbook on stock assessment (Hilborn and Walters 1991). While there's an abundant literature on estimating and forecasting fish stocks, the following case study provides a useful example of how a wide range of different data sources can be fused (chapter 9) in a Hierarchical Bayes framework to improve predictions.

## 10.2 CASE STUDY: BALTIC SALMON

The Baltic Sea is a brackish sea with a long salinity gradient, ranging from 1–2 parts per thousand in the north to around 20 parts per thousand around the entrance to the North Sea. This gradient supports a wide range of genetically unique species or populations. The Baltic salmon is a geographically isolated and genetically unique subpopulation of Atlantic salmon (*Salmo salar* L.) found in the Baltic. Baltic salmon can be subdivided further into different stocks based on their different spawning rivers. Like many other fisheries, the Baltic salmon experienced a period of decline due to overfishing and the damming of spawning rivers, with population lows occurring, in this case, in the 1980s. Fortunately, long-term sustainability became an explicit management goal under the 1997 Salmon Action Plan (HELCOM and IBSFC 1999), and the Baltic salmon stocks are now managed under the EU Common Fisheries Policy (CEC 2009). These fisheries have recovered considerably, but management remains challenging due to disease (M74 syndrome outbursts can decimate first-year cohorts), pollutants (for example, toxic levels of dioxins), climate change, and other regional anthropogenic and natural impacts.

Like much of fisheries management, stock assessments of Baltic salmon in the 1990s relied on deterministic population models implemented in spreadsheets. However, these projections were often very sensitive to parameter choices, so in 1997 researchers in the area began incorporating Bayesian belief networks as part of generating fishery quota recommendations (Varis and Kuikka 1997). The Bayesian approach was progressively extended to incorporate different data sets and assess different parts of the salmon life cycle, the synthesis of which is presented in Michielsens et al. (2008) and reviewed by Kuikka et al. (2014). The current case study is largely derived from these syntheses.

The core of the Baltic salmon stock model (figure 10.2) is an age-structured state-space population model (F). This model takes the same structure as classic matrix population models (see equation 7.1), using it as the process model in a state-space time series model (chapter 8). The outputs of this model are state estimates of the historical and current stock, harvest, recruitment, maturation, and mortality for the different subpopulations, as well as predictions for these quantities, and an overall assessment of the probability of reaching management goals under alternative actions. This model is fundamentally a mark-recapture model based on raw input data on tagging and subsequent catch, which is assumed to follow a Negative Binomial

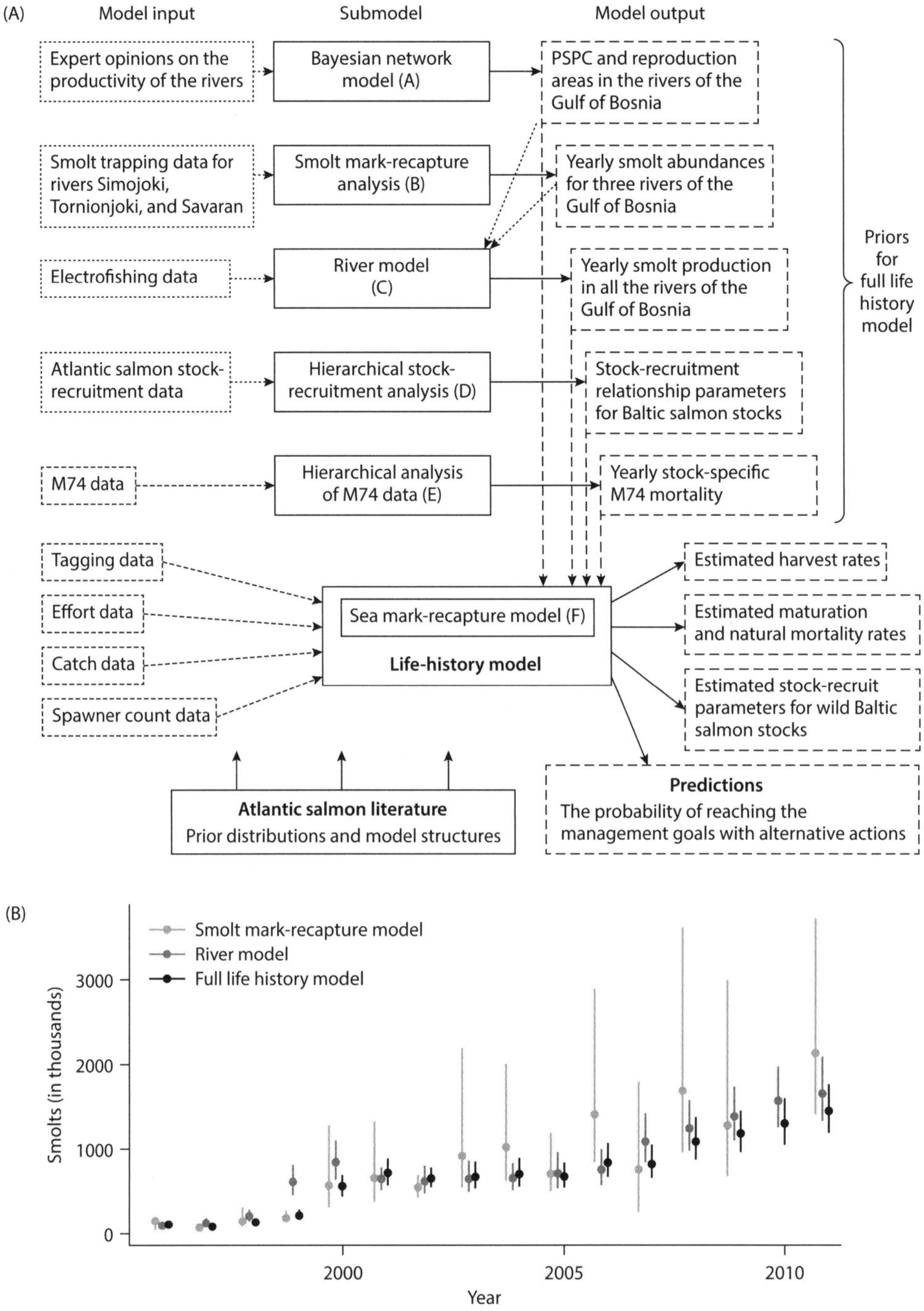

FIGURE 10.2. Baltic salmon stock model. Top: General workflow of the overall data fusion that illustrates the data sets used (Model Input), the process that was estimated (Submodel), and the output of each submodel (Model Output) that then feeds into the overall model (F) as informative priors. Bottom: Impact of sequentially combining data on population stock estimates through stages A, B, and F. Confidence increased at each stage, with the most dramatic change coming from adding the river model data. Reproduced from Kuikka et al. (2014).

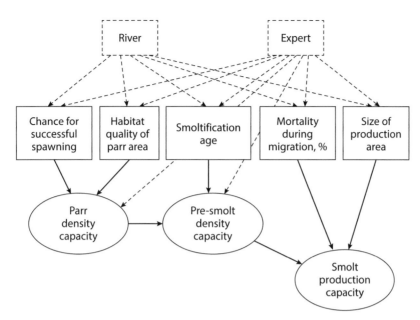

FIGURE 10.3. Bayesian network model for eliciting expert opinion about Baltic salmon smolt production capacity on a river-by-river basis. Reproduced from Uusitalo et al. (2005).

distribution, as well as data on fishing effort (Michielsens et al. 2006b). In addition, the model incorporates age composition data on fish spawning in rivers, which is assumed to follow a beta-binomial distribution (Michielsens et al. 2008). Both of these data models were selected because they are overdispersed forms of the more common Poisson and Binomial models for count data (section 6.1).

What sets the Baltic salmon model apart, and makes it a good example of data fusion (chapter 9), is the number of detailed analyses that feed information into the core model. Specifically, these analyses construct highly informative priors based on a synthesis of the Atlantic salmon literature and the outputs of a range of submodels used to estimate specific demographic parameters (figure 10.2, A–E). All these different submodels and data sets could in theory be incorporated directly into the final model, but instead a multistage approach was chosen for computational reasons. The multistage approach also has practical benefits, in that it allows the outputs of submodels to be independently assessed, rather than trying to diagnose the behavior of the entire model at once.

The first submodel (A) uses expert opinion to assess the *potential capacity* of different rivers to produce smolts (juvenile fish ready to leave the river) using a Bayesian network model (Uusitalo et al. 2005) to combine a number of conditional probability statements into an overall capacity estimate (figure 10.3). In this model two variables, *chance for successful spawning* and *habitat quality of parr area,* are used to estimate the *parr density capacity,* where *parr* are the juvenile freshwater salmon that remain in the river for 2 to 6 years. Next, *smoltification age* and *parr density capacity* are used to estimate the *presmolt density capacity.* Finally, the *presmolt density capacity* is combined with the *size of production area* and *% mortality during* *migration* to estimate the overall *smolt production capacity*. The model also contains two auxiliary variables, *river* and *expert*, as the analysis was performed on ten rivers with the probability distributions for all parameters elicited from five experts (see box 5.2). The general approach employed here, of breaking down what is a conceptually difficult question (how many smolts can be produced by each river) into a series of smaller questions that can be estimated more easily, is a useful example of Fermi estimation—a "back of the envelope" approach to approximation popularized in physics and engineering by Enrico Fermi that has been used to improve forecast skill across a wide range of disciplines (Tetlock and Gardner 2015). Even though there are a large number of parameters involved in this model, because each part is more easily assessed than the whole, this approach can produce more informative estimates than simply eliciting experts about the final smolt capacity directly.

The second submodel (B) is a mark-recapture model used to estimate the annual abundance of migrating smolt (Mäntyniemi and Romakkaniemi 2002). As with the core model, a beta-binomial model is assumed to account for the overdispersion resulting from the schooling behavior of fish. The daily catch probability and the travel time from the site of release to the site of recapture are modeled as logistic and lognormal (respectively) functions of water level and temperature, while the overall recapture over multiple days of trapping follows a multinomial-Dirichlet model (the multivariate generalization of the beta-binomial). Smolt recapture data is only available for three rivers, the Savaran, Simojoki, and Tornionjoki, so the posteriors from this analysis enter as informative priors just for those three rivers.

To estimate smolt abundance for the remaining rivers a third submodel (C) was constructed that fit a hierarchical linear regression between electrofishing data on parr abundance and the mark-recapture estimates of smolt production. This relationship was then used to predict the smolt production in rivers that had parr electrofishing data but no smolt mark-recapture (Michielsens et al. 2008).

The fourth submodel (D) aims to provide informative prior parameter estimates to the density-dependent Beverton-Holt relationship between egg production and the number of recruits (Michielsens and McAllister 2004). No empirical data on this relationship exist for Baltic salmon and therefore a hierarchical fit of the Beverton-Holt was performed across data from nine other Atlantic salmon populations from Canada, Iceland, France, and the UK. Furthermore, the Beverton-Holt model was reparameterized to specifically estimate the steepness of the recruitment relationship, under the assumption that the magnitude of recruitment will be more variable from site-to-site than the curvature of the relationship. Overall this steepness was found to be relatively consistent among sites (CV 23%).

A final submodel (E) was then developed to adjust the stock-recruitment functions from (D) to account for M74 syndrome mortality (Michielsens et al. 2006a). This analysis combines detailed field data on morality for a subset of rivers (Tornionjoki and Simojoki) with lab-based egg incubations in a hierarchical model to allow annual stock-specific predictions of mortality for rivers with complete observations (both prevalence of M74 in spawning females and lab incubations), partial observations (field prevalence *or* lab incubations), or no observations. M74 mortality was shown to have high interannual variability, fluctuating from close to 100% to less than 20%.

Overall, this case study demonstrated that information from a wide range of field observations, along with literature data, expert opinion, and lab incubations, could be effectively combined and that each additional layer of information serves to further constrain stock estimates (figure 10.2, lower panel). Furthermore, a recurring theme in the submodels was the ability to use a hierarchical model structure to leverage information from one stock to help constrain others (that is, to borrow strength across rivers and stocks), while formally accounting for the additional variability and greater uncertainty associated with doing so (relationships from one system were not just blindly applied to others).

This case study also illustrates a number of challenges faced by such large, synthetic analyses. As noted earlier, this model was fit sequentially, rather than all at once, due to computational limitations. Doing so comes at the cost of some loss of precision, which is particularly likely to show up in the covariances (that is, underestimation of trade-offs and feedbacks because different parts were fit separately). In their review Kuikka et al. (2014) also explicitly comment on the time spent on technical problems and MCMC convergence, which led to challenges in timing and co-ordination when pulling together large multipart analyses. Computational cost also limited the ability to use these models in "real-time" to explore different scenarios and policy choices with decision makers (chapter 17), something that might be approximated through simplified models or emulators. As discussed in chapter 4, there are many informatic and workflow challenges in bringing an ecological forecast to an operational state. The computational demand of refitting the whole model every year as new data becomes available is one of the arguments for using sequential approaches to data assimilation, which will be introduced in chapters 13 and 14. Finally, while not presented in this case study, Kuikka et al. (2014) also comment on the critical importance of sensitivity and uncertainty analyses (chapter 10) and openness (chapters 3 and 4), which are particularly important when working on ecological forecasts that address contentious issues.

## 10.3 KEY CONCEPTS

1. Biotic natural resources forecasts have a deep history. In particular, the latent nature of fishery stocks has driven both the early adoption and current sophistication of these forecasts.
2. The Baltic salmon case study illustrated the ability to fuse a remarkably wide range of data sources (chapter 9), but also highlights the challenges of putting complex workflows into practice (chapter 4).
3. The hierarchical structures in the Baltic salmon submodels were able to borrow strength across rivers and stocks, allowing well-studied rivers to constrain others, while formally accounting for the additional variability and greater uncertainty associated with doing so.

# 11

## Propagating, Analyzing, and Reducing Uncertainty

*SYNOPSIS: This chapter focuses on tools and techniques used to propagate uncertainties into forecasts, diagnose what's driving the uncertainty in a model, and use these insights quantitatively to help with the design of new experiments*

IF THE GOAL in chapters 5 and 6 was to understand how to quantify different sources of uncertainty, the goal of this chapter is to understand how those sources of uncertainty affect our forecasts. Consider the general case of a model that predicts a response, $Y$, and we are interested in how a factor, $X$, affects that response, where $X$ might be an input, parameter, initial condition, or any of the other sources of uncertainty we've discussed previously. We will begin with the Sensitivity Analysis question, "How does a change in $X$ translate into a change in $Y$?" We will then build upon this to ask the Uncertainty Propagation question, "How does the uncertainty in $X$ affect the uncertainty in $Y$?" Not surprisingly, the outcome of such an analysis depends upon two factors, how sensitive $Y$ is to $X$ and how uncertain we are in $X$. Next we will consider the case where we have more than one source of uncertainty and we want to ask the Uncertainty Analysis question, "Which sources of uncertainty are most important?" Finally, we will consider a number of tools for model-data feedbacks that aim to answer the Optimal Design question, "How do we best reduce the uncertainty in our forecast?" In every section of this chapter we will introduce a variety of methods to address each question, because within each there are common trade-offs among different methods, such as analytical derivation versus numerical approximation, computational costs, and whether the approach returns the full probability distribution or just the mean and variance.

### 11.1 SENSITIVITY ANALYSIS

The goal of a sensitivity analysis is to understand how a change in $X$ translates into change in $Y$. In most cases the $X$ of interest is a continuous variable, such as the value of a model input, initial condition, or parameter. However, $X$ can also be a categorical variable, such as a choice between alternative models or discrete scenarios. We'll begin by focusing on continuous variables, though most techniques translate well to categorical variables as well.

For the continuous case, the most common definition of *sensitivity* is the change in $Y = f(X)$ given some change in $X$, which leads naturally to the derivative, $dY/dX$, as our most common measure of model sensitivity. Indeed, for simple models it is generally possible to analytically solve for this derivative. However, for all models except a straight line the value of the sensitivity will change as a function of $X$. If the derivative is evaluated at a specific point, which is typically at the mean of $X$, $\overline{X}$, this is referred to as a *local sensitivity*, since that sensitivity only applies to that one location. Local sensitivities have the advantages of being easy to compute and interpret, and not being dependent on the PDF of $X$.

It should be noted that a sensitivity has units of $Y/X$, and thus it is not always straightforward to compare the sensitivity of a model to different variables. One way to solve this units problem is to standardize the units by the mean, by multiplying the sensitivities by $\overline{X}/\overline{Y}$, which leaves a dimensionless quantity known as an *elasticity*. An elasticity also has an intuitive interpretation, where an elasticity of 1 means that a unit change in $X$ results in an equivalent unit change in $Y$. For example, if you increase $X$ by 10%, you also will increase $Y$ by 10%. Similarly, an elasticity greater than 1 means that the response variable $Y$ increases disproportionately to a change in $X$. It should also be noted that since a sensitivity is a slope, both sensitivities and elasticities can be negative. For example, an elasticity of $-0.5$ means that a 100% increase in $X$ will lead to a 50% decrease in $Y$.

For complex models it is often impossible or impractical to analytically calculate a derivative. In this case, it is common to estimate the local sensitivity by *numerical approximation*. There are a number of numerical methods available for estimating derivatives, the simplest being

$$\frac{df}{dx} \approx \frac{f(\overline{x} + h) - f(\overline{x})}{h} \tag{11.1}$$

which, in the limit as $h$ goes to zero, is in fact the classic definition of a derivative. However, due to numerical rounding issues, $h$ should be considerably larger than the machine precision, $\varepsilon$, and in fact the optimal $h$ is approximately $\sqrt{\varepsilon}x$ (where $\varepsilon = 2.2 \times 10^{-16}$ on a typical 64-bit machine) (Press et al. 2007). More precise methods exist for numerical derivatives, but the cost of increased precision is the need for additional evaluations of the function, $f(x)$. However, in practice, the goal of a sensitivity analysis isn't to get the best possible estimate of the derivative, but to understand the response to realistic variation, and thus a larger $h$ is frequently used, such as $\pm 10\%$ or 20% of $\overline{X}$.

When considering models with multiple parameters, the simplest approach to sensitivity analysis, often referred to as the *one-at-a-time (OAT)* sensitivity analysis, involves holding all parameters at their mean except one, and then numerically evaluating the sensitivity of the model (equation 11.1) to that one variable (figure 11.1). That variable is then returned to its mean and the process is repeated for the next variable. However, the OAT approach faces two criticisms, both of which are associated with the local nature of the approximation. The first is that, as noted earlier, sensitivity is not constant across parameter space unless the response is a straight line. The second is that, when considering more than one variable, the OAT approach ignores parameter interactions.

The alternatives to local analyses, termed *global sensitivity analyses*, require evaluating the model over a wide range of values, including simultaneous perturbations to multiple parameters. The goal of such analyses is to explore a wide (yet realistic) range of parameter values to better understand the overall sensitivity of a model rather than just the local sensitivity at the mean. The more you explore different

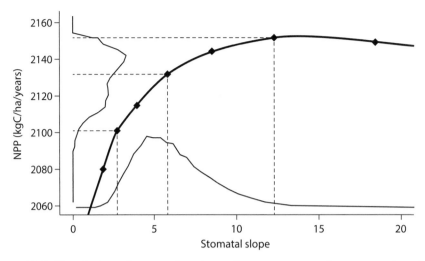

FIGURE 11.1. Translation of uncertainty in an input parameter into uncertainty in an output. The bold black curve illustrates the sensitivity of NPP to stomatal slope, with the black diamonds denoting the median and quantile equivalents of ±1,2,3 standard deviations for stomatal slope based on a one-at-a-time sensitivity analysis. The probability density on the *x*-axis represents the uncertainty in one model input—in this case, the stomatal slope (Leuning 1995) parameter in the Ecosystem Demography terrestrial biosphere model (Medvigy et al. 2009). The probability density on the *y*-axis is the uncertainty in NPP attributable to the uncertainty in stomatal slope, with the dashed lines indicating the translation of specific values. Reproduced from Dietze et al. (2014).

parameter combinations, the better your understand the model, but the more runs you need to do, so in general you want to be a bit strategic in how parameter combinations are chosen. A wide range of options exists for attempting to explore parameter space thoroughly yet efficiently, with the primary trade-off among approaches being that models that are computationally expensive cannot be run as many times and thus cannot explore parameter space as fully. Therefore, the choice of runs must be much more strategic and clever for a complex model, while with simple models it is easier to explore parameter space exhaustively by brute force. The other major trade-off among methods is the size of the parameter space being explored. Because *parameter space increases exponentially with the number of parameters in a model*, evaluating high-dimensional models remains challenging—a problem often referred to as the *curse of dimensionality*. Imagine you've decided to run a one-variable model at 10 different parameter values to evaluate its sensitivity. If the model instead has two parameters, evaluating the 10 × 10 grid of all parameter combinations requires running the model 100 times. A model with 20 parameters, which is by no means unusual in ecology, would then require 100,000,000,000,000,000,000 runs. To put that in perspective, if your model only took 1 second to run, and you started your sensitivity analysis immediately after the Big Bang, you would currently be only 0.4% done.

   It should be noted that most global sensitivity analyses are not actually "global" in their extent, but require that you specify the range over which each parameter is being evaluated. This is worth giving careful thought to since the results of a sensi-

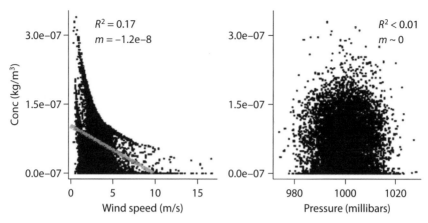

FIGURE 11.2. Monte Carlo sensitivity analysis of Gaussian Plume atmospheric dispersion model to variability in wind speed and atmospheric pressure. Sensitivity is approximated by the slope, $m$, while uncertainty partitioning is approximated by the $R^2$.

tivity analysis will themselves be sensitive to the domain chosen. For example, varying all parameters over a range of $\pm 50\%$ may give a different answer from $\pm 10\%$ and neither may be relevant if one parameter varies $\pm 500\%$ while another varies $\pm 1\%$. Because of this, my recommendation, especially if you are going to conduct the uncertainty propagation and analyses described in the next sections, is that *sensitivity analyses be tied explicitly to the probability distribution of the inputs*. For example, the range for a parameter might be set to its 95% or 99% CI (or roughly 2 or 3 standard deviations).

There's a broad range of options for global sensitivity analyses, with whole books on the topic (Saltelli et al. 2008), so here I will just summarize a few options that are particularly useful or common. The simplest, most brute-force approach is a *Monte Carlo (MC) sensitivity analysis,* in which the $X$ values are simply sampled from their joint posterior distribution and the model is run for each sample. When a data-informed posterior is not available you must instead rely on your prior estimates for the inputs—the use of uniform prior distributions has been particularly common, though not necessarily recommended (chapter 5). The number of runs required depends on the dimensionality, speed, and nonlinearity of the model: models with large numbers of parameters, or which are highly nonlinear, require more runs; fast models allow more runs to be done in the same amount of time. In addition to its simplicity, the Monte Carlo approach has the advantage that this exact same analysis shows up in the following sections on error propagation and uncertainty analysis and all these analyses can be done from the same set of runs. To calculate sensitivity from the Monte Carlo output, the simplest approach is to fit a multiple regression model to the model output as a function of the randomly sampled inputs (figure 11.2). The regression slopes for each variable are the global sensitivities, and interactions between different $X$'s can be assessed by including parameter interaction terms in the regression model. Furthermore, the $R^2$ of each linear regression provides an estimate of the relative importance of different input variables. Scatterplots are also a critical diagnostic as linear sensitivities are not always a good approximation. It's straightforward to add polynomial terms to the sensitivity analysis if there's a need to get a more precise $R^2$ estimate or to estimate how sensitivity changes across the input space. In some cases, the goal may explicitly be to build up a more general statistical approximation to a model (box 11.1), in which case more sophisticated and flexible statistical models may be fit (Gaussian process models, multivariate splines, general additive models [GAMs], and so on). That said, in many cases the goal is much more mundane, such as generating a single score to compare parameters, in which case the basic linear regression is perfectly adequate.

As noted earlier, another approach to sensitivity analysis is to use an *emulator*, which is a statistical model that approximates the output of the model being investigated (see box 11.1). Emulators are typically used for models that have high computational costs, where brute-force methods are not feasible. In this case the emulator is used *in place of* the actual model in more detailed analyses, such as the MC approach. As we'll see later, an emulator can also be useful for uncertainty analysis, error propagation, and data assimilation.

For computationally demanding models there are alternative approaches that sample parameter space more strategically. One such approach is the *Elementary Effects (EE)* method, which takes the mean of many OAT sensitivities at $r$ locations distributed across parameter space rather than just centered on $\overline{X}$. We define an elementary effect, $EE_i^j$, as the OAT sensitivity (equation 11.1) of the $i^{th}$ input centered around the $j^{th}$ location in input space. If we evaluate a total of $k$ model inputs at a total of $r$ locations in input space, then our EE sensitivity for input $i$ is its mean elementary effect:

$$\mu_i^* = \frac{1}{r} \sum_{j=1}^{r} |EE_i^j|$$

In addition, with the EE method one can also calculate a standard deviation among sensitivities at the $r$ locations in input space, which indicates how variable the sensitivity of a parameter is. For a linear model this SD would be zero, since it has the same sensitivity everywhere, but for other models higher SD would indicate greater variability in sensitivity. Within the EE method there are a variety of sampling strategies that have been proposed to choose the $r$ locations, but in most versions this approach takes $r(k + 1)$ model evaluations.

In most global sensitivity analyses, such as the EE and emulator methods, one common question is how to sample parameter values from their distributions. With the MC approach we relied on random samples, but this is an inefficient approach, as it can lead to clumped samples in some parts of parameter space and voids in others. More commonly, parameter sampling is stratified in some way to ensure well-dispersed samples; however, as discussed earlier, uniform sampling in each dimension is computationally prohibitive for most models. A number of other sampling approaches have been proposed, such as the Halton sequence, Sobol's LP sequence, and the Latin Hypercube (LHC), each of which produces a multivariate sample that is *marginally* approximately uniform for each parameter and space-filling in multiple dimensions (Saltelli et al. 2008). Of these approaches the LHC is the most common and simple, as it involves generating a uniform sequence of values for each parameter and then randomly permuting the order of each parameter independently. In all approaches, for posterior distributions that are nonuniform it is customary to sample the quantiles of the distribution so that the cumulative distribution is approximately uniformly sampled.

## Box 11.1. Emulators

What do you do if your model is too computationally expensive to perform the sorts of analyses described in this book? This can be a common problem for computer simulation models, where individual model runs may take hours to days to complete. While one option is to build a simpler model, another is to build a statistical model, known as an *emulator*, that approximates the complex model. An emulator is not a first-principles approximation to the model itself. Instead, a set of model runs is required to construct an emulator and the relationship between model inputs and outputs is inferred by treating these like data. However, the number of runs required to build an emulator is far fewer than for Monte Carlo methods (MC sensitivity and error propagation, MCMC calibration). Furthermore, this set of runs can be run simultaneously (for example, on different nodes on a cluster), which makes the problem naïvely parallelizable. In practice, this can provide another huge speed-up. For example, one might perform a few hundred model runs, each on a different computer, while MCMC might require >10,000 runs done sequentially.

The basic steps for working with an emulator are:

1. Construct a statistical design of how the inputs are sampled.
2. Run the model for this set of inputs.
3. Fit the emulator (statistical approximation).
4. Perform analyses using the emulator (sensitivity, uncertainty, MCMC, and so on) the same as if it were any other model.

Step 1: When constructing the statistical design, you don't want to sample randomly, as that results in some parts of parameter space being too clumped and others too sparse. You also don't want to sample systematically because of the curse of dimensionality. Rather, statistical designs are used that are both space filling and efficient, such as a Latin Hypercube or Halton sequence (see section 11.1). As with the sensitivity analysis, sampling parameter space based on the quantiles of the posterior distribution generally works better than sampling parameters uniformly, as this better captures both the center of the posterior and the tails.

Step 3: Unlike the MC sensitivity analysis, where we fit a regression through the point cloud, with an emulator we want the approximation to return the *exact* correct result if we pass it the original design points. Therefore, we want to choose statistical methods to interpolate between points in parameter space. The classic approach for this is to use a *Gaussian process model* (Sacks et al. 1989), which is analogous to *Kriging* (interpolation based on spatial autocorrelation, chapter 6) in $n$-dimensional parameter space. However, other approaches such as multivariate splines, general additive models (GAMs), and other machine-learning algorithms (chapter 16) have been used. See the following papers for more information about emulators (O'Hagan 2006); their diagnosis (Bastos and O'Hagan 2009); and their use in sensitivity analysis (Oakley and O'Hagan 2004), uncertainty analysis (O'Hagan et al. 1998; Kennedy et al. 2006), and calibration (Kennedy and O'Hagan 2001).

Another technique for global sensitivity analysis is to vary multiple parameters at once, an approach called *group sampling*. For very computationally costly or high-dimensional models it is even possible to estimate sensitivities using less than $k$ model runs using what is known as a *supersaturated design*. This approach is based on the assumption that most parameters have a small impact on model output and that, at least initially, the goal is to identify important parameters more than to quantify the effect size of every parameter. For example, if $k = 128$ parameters, the average parameter contributes <1% to the observed variability in model outputs. Typically, most parameters contribute very little, while a small number of influential parameters contribute substantially more. For the sake of this example, let's imagine that 4 parameters are important. Instead of evaluating each parameter using an OAT design, we might divide the parameters randomly into 16 groups of 8 parameters each. We would then do 16 model runs, and in each run we perturb all 8 parameters within a group at the same time. From these runs we will find that most groups produce similar outputs, since they contain only noninfluential parameters, but a few groups will be identified as containing important parameters. At this stage we cannot identify which parameters are the important ones in those groups versus which are false positives. Nor can we rule out that another group contains important parameters that are canceling each other out (false negatives). Therefore, in the next round of model runs we reshuffle which parameters are in which groups. This could be done randomly, or we could be more systematic to make sure that parameters from the influential groups are divided into separate groups. If there were two important parameters that were canceling each other out in the first round, there is now only a 1/16 chance that they are in the same group again. Figure 11.3 shows a simulated example of evaluating a model with this design, where in each iteration the group containing the 4 sensitive parameters was clearly identified. The intersection of these

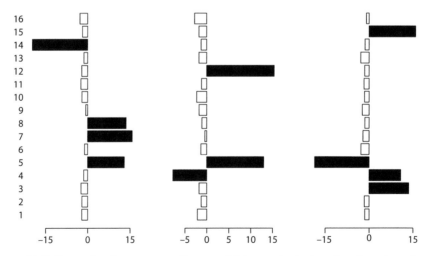

FIGURE 11.3. Example of group sampling sensitivity analysis showing three iterations of 16 groups (48 runs total) evaluating a 128-parameter model (8 parameters per group). The *x*-axis is group sensitivity, and the *y*-axis is group number. Black bars indicate parameter blocks that contain highly sensitive parameters.

sets labels a parameter as having been selected 0, 1, 2, or 3 times. As the number of iterations increases, the odds of an important parameter being identified zero times drops rapidly, as does the number of false positives being identified in all iterations. What can be more challenging is balancing false positives versus false negatives for parameters scored an intermediate number of times (in this case 1 or 2 times). Additional tricks exist for improving group fidelity, such as changing the number of groups among iterations, dropping parameters that have been identified with high probability, and randomly sampling the sign of the perturbations (Saltelli et al. 2008).

Finally, there are a number of variance-based methods for estimating global sensitivities. I will discuss these methods in section 11.3 because if performed on a data-informed posterior, or well-constructed prior, these are actually uncertainty analyses, because they allow you to make direct inferences about how the uncertainties in your inputs control the uncertainties in your outputs. However, if performed by sampling from broad uniform input distributions, as is often done, it is more appropriate to consider these methods to be sensitivity analyses. This illustrates a key but subtle point, *that how you treat uncertainties changes the nature and meaning of many analyses*, and just knowing the name of a tool or approach is often insufficient to interpret the results. In this case variance-based sensitivities swap from being an extremely useful tool for diagnosing uncertainties, to being hard to interpret, depending on whether the input sampling distributions represent the true uncertainty in the parameters or arbitrary choices.

## 11.2 UNCERTAINTY PROPAGATION

Uncertainty propagation refers to the process of translating uncertainty in our model's inputs, $X$, or parameters, $\theta$, into uncertainty in the model's outputs, $Y$ (figure 11.1). This is an increasingly important goal in ecological analyses and a fundamental part of ecological forecasting—as has been stated throughout this book, a forecast without a statement about its uncertainty is of limited value. A common example of uncertainty propagation is a regression confidence interval, which is generated by propagating the regression parameter uncertainty (slope and intercept) through the regression model to estimate the uncertainty in the output ($Y$). Even outside of modeling, uncertainty propagation is increasingly demanded when generating data or working with data generated by others. For example, something as simple as reporting a temperature involves more than just translating the millivolts recorded by a sensor into Kelvin, but also involves accounting for the uncertainty in the sensor calibration curve, the sampling uncertainty across sensors, and the increase in uncertainty across time due to sensor drift (chapter 9).

At their essence, all methods for uncertainty propagation are trying to do the same thing—translate the probability distribution of $X$ into the probability distribution of $Y$ through some function $y = f(x)$. As such, *all approaches depend upon knowing the uncertainty in X and the sensitivity of Y to X*. In this section we will discuss a number of methods for uncertainty propagation (table 11.1), which are organized based on approach (analytical solution or numerical approximation) and output (full probability distribution or statistical moments—typically mean and variance). Not surprisingly, the methods for returning the full PDF are more costly than those that return only moments, either in terms of computational time (for numerical methods) or

TABLE 11.1. Methods for Uncertainty Propagation

| | Output | |
| Approach | Distribution | Moments |
| --- | --- | --- |
| Analytic | Variable Transform | Analytical Moments |
| Numeric | Monte Carlo | Taylor Series Ensemble |

mathematical complexity (for analytical methods). Along the other axis, analytical methods typically provide a more general and insightful solution than a numerical approximation, but the difficulty of implementing analytical methods increases quickly as models become more complex. By comparison, numerical methods tend to be simple to implement regardless of model complexity.

### 11.2.1 Variable Transform

Let's begin our discussion of uncertainty propagation with the analytical transformation as our "gold standard," since it provides an exact analytical solution for the full PDF. In this problem our input $X$ is a random variable described by the PDF $p_X[x]$ and our response variable $y$ is predicted by the model $y = f(x)$. Our goal is to solve for the full PDF of the random variable $Y$, $p_Y[y]$ given the input $X$. This has a general analytical solution

$$p_Y[y] = p_X[f^{-1}(y)] \left| \frac{df^{-1}(y)}{dy} \right|$$

the proof for which can be found in most textbooks on probability (Casella and Berger 2001). The first term of this solution involves analytically solving for the inverse of the model, $x = f^{-1}(y)$, and substituting that into $p_X[x]$. The second term involves solving for the derivative of that inverse and then taking its absolute value. In the multivariate case $p_X$ becomes a multivariate PDF, $f$ is a multivariate model that needs to be inverted, and the final term becomes the Jacobian matrix of partial derivatives of that inverse.

The challenges to this approach are that many models don't have an analytically tractable inverse, the Jacobian will be tedious to solve, and even if all that is done successfully it is unlikely that the resulting PDF will match a known, named distribution. If your solution is not a known distribution, you won't know how to interpret the PDFs in terms of common statistics such as the mean and variance, or how these statistics are related to the parameters of the model. Therefore, you will be faced with having the additional task of analytically solving for the mean, variance, and any other summary statistic you need:

$$E[y] = \int_{-\infty}^{\infty} y \, p_Y[y] dy$$
$$Var[y] = \int_{-\infty}^{\infty} (y - E[y])^2 p_Y[y] dy$$

Even if it is possible to solve $p_Y[y]$ analytically, these moment integrals may be difficult or impossible to solve. All of this is to say that, on a practical basis, you are unlikely to encounter or employ this "gold standard" for all but the simplest cases where you know, a priori, that the final result will be a named distribution.

### 11.2.2 Analytic Moments

An analytical transformation of a full PDF is often challenging. By contrast, it is often much simpler to transform just the moments of a distribution using known properties of statistical moments. In particular, we are frequently most interested in the mean and variance of Y. For linear models the statistical properties of means and variances (appendix A) can be applied to give an *exact* calculation of $E[Y]$ and $Var[Y]$ given $E[X]$, $Var[X]$, and $f(X)$. In applying these properties the key is to apply them to the model from the outside inward, similar to what one does when taking derivatives, and not to forget about covariances or Jensen's inequality. Consider, as an example, the simple linear model $y = \beta_0 + \beta_1 x + \varepsilon$, where $\varepsilon \sim N(0,\sigma^2)$ and we assume $x$ is known but the intercept, slope, and residual error are all estimated. First we want to calculate $E[y]$:

$$E[y] = E[\beta_0 + \beta_1 x + \varepsilon]$$

Working from the outside, we can apply the property that means are additive, $E[\sum X] = \sum E[X]$

$$E[y] = E[\beta_0] + E[\beta_1 x] + E[\varepsilon]$$

Next we can simplify the middle term by applying the property $E[cX] = cE[X]$, where $c$ is a constant, noting that in this case $\beta_1$ is the random variable and $x$ is the constant. We can also simplify the last term using the properties of a Normal distribution, $E[N(\mu,\sigma^2)] = \mu$, where in this case $\mu = 0$ (residual error is unbiased).

$$E[y] = E[\beta_0] + xE[\beta_1]$$

We can take a similar approach with the variance

$$Var[y] = Var[\beta_0 + \beta_1 x + \varepsilon]$$

where the variance of a sum follows the property that

$$Var\left[\sum X\right] = \sum_i \sum_j Cov[X_i, X_j] = \sum_i Var[X_i] + 2\sum_{i \neq j} Cov[X_i, X_j]$$

Applying this to our model gives

$$Var[y] = Var[\beta_0] + Var[\beta_1 x] + Var[\varepsilon] + 2Cov[\beta_0,\beta_1] + 2Cov[\beta_0,\varepsilon] + 2Cov[\beta_1,\varepsilon]$$

If we assume [and check] that the residuals are homoskedastic and independent, then they do not covary with the slope and intercept, and thus we can drop the last two terms. By contrast, in a regression there is usually a strong covariance between the slope and intercept so the $2Cov[B_0,B_1]$ term cannot be ignored. Similar to the mean, we can also apply the properties of a Normal distribution, $Var[\varepsilon] = Var[N(\mu,\sigma^2)] = \sigma^2$. Finally, we can apply the property for a variance multiplied by a constant, $Var[cX] = c^2 Var[X]$, to give the final result

$$Var[y] = Var[\beta_0] + x^2 Var[\beta_1] + \sigma^2 + 2Cov[\beta_0,\beta_1]$$

Indeed, this is the equation used to add the classic hourglass-shaped Predictive Interval to a regression line. The equation for a confidence interval is virtually identical except that a CI considers only parameter error, and thus does not include the residual variance $\sigma^2$.

### 11.2.3 Taylor Series

When applying analytical moments, it will not take long before you encounter a model that, either in its entirety or for some term, cannot be simplified using the rules of means and variances. In these cases the standard *analytical* approximation is to use a Taylor Series to linearize the model to something that mean and variance rules *can* be applied to. For example, we can approximate the mean of a univariate function, $f(x)$, as

$$E[f(x)] \approx E\left[f(a) + \frac{f'(a)}{1!}(x-a) + \frac{f''(a)}{2!}(x-a)^2 + \cdots\right]$$

If we truncate the Taylor Series after the first three terms, evaluate this centered around the mean, $a = \bar{x}$, and apply the rule for the mean of a sum, this becomes

$$E[f(x)] \approx E[f(\bar{x})] + f'(\bar{x})E[(x-\bar{x})] + \frac{1}{2}f''(\bar{x})E[(x-\bar{x})^2]$$

We can further simplify this by noting that the first expected value is a constant, the second is zero, and the third is the definition of variance, which gives

$$E[f(x)] \approx f(\bar{x}) + \frac{1}{2}f''(\bar{x})Var[x] \tag{11.2}$$

In addition to being a useful predictor, this approximation is a nice demonstration of Jensen's inequality. Most ecologists are familiar with the basic form of the inequality, that the mean of a function, $E[f(x)]$ is not equal to the function of the mean, $f(\bar{x})$. However, Jensen's inequality also states, more precisely, that for a convex function, $f''(\bar{x}) > 0$, the mean of the function is larger than the function of the mean. Given that a variance is always positive, our approximation not only demonstrates this but also shows that $f(\bar{x})$ becomes an increasingly biased estimate of the mean as the uncertainty in $X$ increases or as the curvature of the function increases. Finally, equation 11.2 can further be generalized to the case of a multivariate set of $X$'s as

$$E[f(x)] \approx f(\bar{x}) + \frac{1}{2}\sum_i\sum_j \frac{\partial^2 f}{\partial x_i \partial x_j} Cov[x_i, x_j] \tag{11.3}$$

the form of which is fairly intuitive given the rule for the variance of a sum.

We can derive a similar Taylor Series approximation for the variance of a univariate function as

$$Var[f(x)] \approx Var\left[f(a) + \frac{f'(a)}{1!}(x-a) + \cdots\right]$$

Again we evaluate this at $a = \bar{x}$, though in this case we will only evaluate the first two terms of the Taylor Series. Expanding out the parentheses in the second term this gives

$$Var[f(x)] \approx Var[f(\bar{x}) + f'(\bar{x})x + f'(\bar{x})\bar{x}]$$

Applying the rule for the variance of a sum, we can recognize that $\bar{x}$, $f(\bar{x})$, and $f'(\bar{x})$ are all constants with respect to $x$. Because the variance of a constant is zero the first and third terms are both zero leaving only

$$Var[f(x)] \approx Var[f'(\bar{x})x]$$

Finally, we can apply the property for a variance multiplied by a constant to reach our final approximation

$$Var[f(x)] \approx f'(\overline{x})^2 Var[x] \qquad (11.4)$$

It is worth noting here that the first term, $f'(\overline{x})$, is our analytical definition of sensitivity (section 11.1). If we take the square-root of this function, it simplifies to $SD[f(x)] \approx f'(x)SD[x]$, where SD is standard deviation. Therefore equation 11.4 literally states that *the uncertainty in a prediction is the product of the uncertainty in the inputs and the model's sensitivity to those inputs.* Finally, as with the mean, we can generalize this result to the multivariate case

$$Var[f(x)] \approx \sum_i \sum_j \frac{\partial f}{\partial x_i} \frac{\partial f}{\partial x_j} Cov[x_i, x_j] \qquad (11.5)$$

To demonstrate the Taylor Series approximation, consider the frequently encountered Michaelis-Menten (M-M) equation, and assume that $x$ is known but that the asymptote, $V$, and half-saturation, $k$, have uncertainty

$$y = \frac{V \cdot x}{k + x}$$

Applying equations 11.3 and 11.5 requires that we calculate the first and second derivatives. Recall that in this example we are considering $x$ to be fixed but $V$ and $k$ to be uncertain, which gives us the following first and second derivatives:

$$\frac{\partial f}{\partial V} = \frac{x}{k + x}$$

$$\frac{\partial^2 f}{\partial V^2} = 0$$

$$\frac{\partial f}{\partial k} = \frac{-Vx}{(k + x)^2}$$

$$\frac{\partial^2 f}{\partial k^2} = \frac{2Vx}{(k + x)^3}$$

$$\frac{\partial^2 f}{\partial k \, \partial V} = \frac{-x}{(k + x)^2}$$

Let's start by looking at the univariate predictors, considering just $V$ and $k$ individually. Applying equations 11.2 and 11.4 to $V$ gives

$$E[f_V(x)] \approx \frac{\overline{V}x}{k + x} \qquad (11.6)$$

$$Var[f_V(x)] \approx \frac{x^2}{(k + x)^2} Var[V] \qquad (11.7)$$

where the second term in the mean is zero because the second derivative is zero, which occurs because the relationship between $V$ and $y$ is linear. The equation for the variance implies that the uncertainty in $y$ associated with $V$ is initially zero at $x = 0$

but increases asymptotically to Var[$y$] = Var[$V$], which makes sense since $V$ controls the asymptote of the function and should have less impact as you move away from the asymptote. The top panel in figure 11.4 illustrates the application of these approximations for the parameter $V$ at $x = 2$ assuming that $V$ has a mean of 100, $k$ has a mean of 5, both have a variance of 25, and they have a covariance of –15. In this panel the solid diagonal line illustrates the sensitivity of $y$ to $V$, and the PDF on the $x$-axis is a Normal distribution corresponding to the specified uncertainty in $V$. The PDF on the $y$-axis represents a Normal distribution with the mean and variance

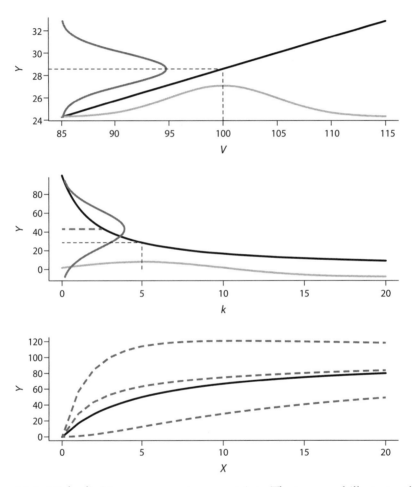

FIGURE 11.4. Michaelis-Menton parameter uncertainty. The top panel illustrates the effect of uncertainty in $V$ alone (PDF on the $x$-axis) on $Y$ (PDF on the $y$-axis) given the sensitivity of $Y$ to $V$ (solid line). Because the sensitivity is linear, the mean of $Y$ is predicted by the mean of $V$ (light dashed line). The middle panel depicts the effect of uncertainty in $k$ on $Y$, which has a nonlinear sensitivity and thus the mean of $Y$ (heavy dashed line) is greater than predicted by the mean of $k$. The bottom panel shows the confidence interval (upper and lower dashed lines) in the relationship between $X$ and $Y$. The mean prediction (center dashed line) is slightly higher than the prediction using the mean parameter values (solid line) due to Jensen's inequality.

from equations 11.6 and 11.7, while the dashed line maps the mean of $V$ to the mean of $Y$.

If we apply equations 11.2 and 11.4 to $k$, this gives

$$E[f_k(X)] \approx \frac{Vx}{\overline{k} + x} + \frac{Vx}{(\overline{k} + x)^3} \, Var[k] \tag{11.8}$$

$$Var[f_k(X)] \approx \frac{V^2x^2}{(\overline{k} + x)^4} \, Var[k] \tag{11.9}$$

As can be seen in the middle panel of figure 11.4 the relationship between $k$ and $y$ is concave u,p which causes $E[y]$ to be slightly larger than $f(x|\overline{k})$. The variance approaches zero at both $x = 0$ and in the limit as $x$ increases and thus there is an intermediate value where the variance is maximized, which can be shown to occur at $x = \overline{k}$. Finally, we can put these responses together by applying equations 11.3 and 11.5, which includes all the terms from equations 11.6 to 11.9 as well an additional term where the covariance reduces the mean and variance.

$$E[f_{kV}(X)] \approx \frac{\overline{V}x}{\overline{k} + x} + \frac{\overline{V}x}{(\overline{k} + x)^3} \, Var[k] - \frac{x}{(\overline{k} + x)^2} \, Cov[V,k] \tag{11.10}$$

$$Var[f_{kV}(X)] \approx \frac{x^2}{(k + x)^2} \, Var[V] + \frac{\overline{V}^2x^2}{(\overline{k} + x)^4} \, Var[k] - \frac{\overline{V}x^2}{(\overline{k} + x)^3} \, Cov[V,k] \tag{11.11}$$

In applying the Taylor Series approximation for the mean and variance it is also useful to note that one can often leverage known approximations to common functions (see appendix B) rather than having to derive the full set of first and second derivatives. In doing so the procedure is much the same as for the analytical moments (section 11.2.2), where you have to work inward from the mean and variance of the full model. In addition, it is not uncommon when working with particularly complex models to use only the linear approximation, thus dropping the second term in the mean (equation 11.2) and the need to solve for a matrix of second derivatives.

It is also worth noting that the moment approximations in both sections 11.2.2 and 11.2.3 have built into them assumptions of Normal error and linearity that can often get you into trouble. For example, in our Michaelis-Menton example imagine that we are now interested in the effects of uncertainty in $x$ and are faced with the situation that $x$ must be positive—for the sake of example let's assume that $x$ follows a Gamma distribution with large uncertainty (figure 11.5, solid PDF on the $x$-axis). In this case the linear variance approximation (fine dashed line), assumes that the uncertainty in $x$ is Gaussian (dashed PDF) and thus predicts a Gaussian variance on the $y$ (dashed PDF). In reality the sensitivity is very nonlinear (solid line) and the true PDF of $y$ (solid line) is both highly skewed and bound over a finite range.

A particularly important example of error propagation, especially in the context of forecasting, is the application of the Taylor Series approach to recursive, discrete-time models, which take the general form $x_{t+1} = f(x_t) + \varepsilon_t$, where $f$ is the deterministic process model and $\varepsilon_t$ is the additive process error. This is the same class of dynamic models considered in the state-space model (chapter 8). Discrete-time population models, such as logistic growth, are a familiar example of such models. If we assume that the parameters of $f$ are known and that $\varepsilon_t$ is distributed with a mean of 0 and a

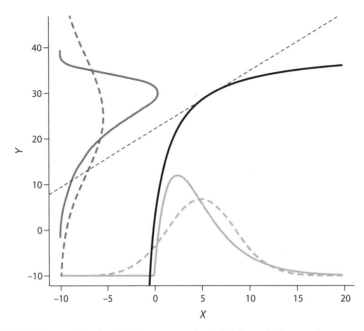

FIGURE 11.5. Failure of Taylor Series approach. Solid lines indicate the true uncertainty in $X$, the true Michaelis-Menton function, and the resulting true uncertainty in $Y$ that results from propagating the uncertainty in $X$ through the function. The dashed lines indicate the Normal approximation to the error in $X$, the linear approximation to the model, and the resulting Normal approximation to the error in $Y$.

variance $q$, then we can apply equations 11.2 and 11.4 to find a general approach for propagating uncertainty through time into most classes of ecological forecasts:

$$E[x_{t+1}] \approx f(\overline{x}_t) + \frac{1}{2} f''(\overline{x}_t) Var[x_t] \qquad (11.12)$$

$$Var[x_{t+1}] \approx f'(\overline{x}_t)^2 Var[x_t] + q \qquad (11.13)$$

From these equations we can see that the uncertainty in our forecast is dependent upon three components: the uncertainty about the state, $Var[x_t]$, the sensitivity of the system, $f'(\overline{x}_t)^2$, and the process error, $q$. The process error always increases uncertainty with each time step, and all else being equal will compound and cause forecast uncertainty to steadily increase moving into the future. On the other hand, if $|f'(\overline{x}_t)|$ < 1, then the deterministic component of the forecast is stabilizing and, in the absence of process error, the uncertainty would gradually dissipate as the system converges toward equilibrium. Indeed, if you were to look back to chapter 2, you'd see that $|f'|$ < 1 is the criteria for whether a population model is stable or unstable. The uncertainty in the overall forecast thus depends upon the relative strength of these two components. By contrast, if $|f'|$ > 1, then the process itself is unstable, and both the deterministic components and the process error are increasing model uncertainty through time. Still, even in this context, these two components may be of very different strengths. For example, in weather forecasting the unstable, chaotic nature of the system dominates over the process error, which is why data assimilation in weather forecasting is predominantly an initial condition problem—to minimize $Var[x_{t+1}]$ you need to minimize $Var[x_t]$, the uncertainty about the current state. By contrast, the supposition throughout this book is that most ecological systems have stabilizing feedbacks and thus it is the process error that dominates forecast uncertainty. *In that case, ecological forecasts are best improved by understanding and partitioning the process error* (chapter 6). More generally, as was discussed in section 2.5.3, the preceding analysis can be extended to consider not just initial condition and process error, but also parameter error, parameter variability (random effects), and covariate/driver error.

### 11.2.4 Monte Carlo (Distribution)

So what do we do in the many cases when an analytical approximation to uncertainty propagation is impossible, impractical, or inaccurate? Fortunately, there exists a numerical approach that is simple, robust, applicable to almost any problem, and provides (an approximation of) the full probability distribution as its output—Monte Carlo simulation. The idea behind Monte Carlo simulation is the same as that discussed in chapter 5, that we can approximate a probability distribution with a random sample of values from that distribution. These samples can be used to approximate the full distribution (histogram) or summarized in terms of various statistical quantities of interest, such as the mean, variance, and quantiles. While Jensen's inequality teaches us that we cannot simply plug statistical moments into a nonlinear function and get the correct answer out, it is perfectly valid to plug individual $X$ values in and get out the correct $Y$. Therefore, it turns out that if we want to transform a probability distribution through a model (which is exactly what uncertainty propagation is), then all we have to do is transform a sample from that distribution. This leads to a very simple algorithm:

1. Sample random values, $x_i$, from the (joint) probability distribution of $X$.
2. Calculate $y_i = f(x_i)$ for all samples of $X$.
3. Use the sample of $y$'s to approximate the PDF of $Y$.

Within this approach there are a few nuances that are worth mentioning. First, if $X$ is multivariate, then it is important to sample from the joint distribution of all $X$'s to account for covariances. Second, if your response of interest is the functional relationship itself, such as the predicted curve from a model or a forecast time series, then each sample is the full response curve. To provide an example, imagine that $f$ is a linear regression and there is uncertainty in the slope and intercept (figure 11.6). We might draw a random [slope, intercept] pair from their joint distribution and then, for a sequence of input $x$ values, predict a sequence of $y$ values along that line. In practical terms, this means that the prediction is often a matrix, with rows corresponding to different iterations of sampling and columns corresponding to predictions for different $X$'s. Each column is a histogram, which might be summarized based on sample quantiles to estimate a confidence interval. It is important to note in this approach that one row (that is, one line) counts as one sample. Sampling different [slope, intercept] pairs independently to make predictions for different $X$ values would negate the substantial covariance between different predicted values, which in many applications may lead to a substantial overestimation of the uncertainty in the process. Furthermore, as we will see in the later chapters on data assimilation, that

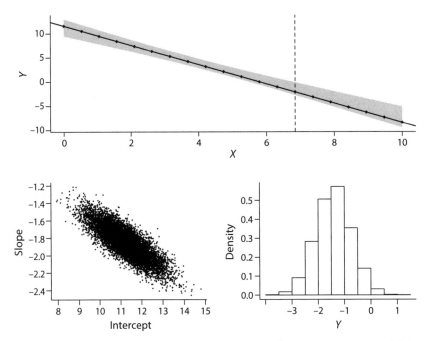

FIGURE 11.6. Monte Carlo uncertainty propagation in a linear regression model. Bottom left: Bivariate scatterplot of samples from the slope and intercept illustrates the strong negative correlation between these parameters. Top: Regression confidence interval (shaded area) and the line drawn from a single sample from the slope/intercept distribution (solid line) evaluated across a sequence of points (diamonds). Bottom right: Histogram of predicted Y values along the dotted vertical line in the top panel.

covariance is often exploited to further constrain predictions when new observations become available.

As with other Monte Carlo approaches, the accuracy of the estimate depends upon the size of the Monte Carlo sample from the input distribution. The sample size needed depends strongly upon the quantities of interest, with the required sample size increasing as you move outward into the tails of the distribution. Thus it takes more samples to get a reliable estimate of a variance than it does for a mean, and more samples to estimate a 95% confidence interval than a variance. In practice, a Monte Carlo simulation size around 5000 is sufficient for most applications to smoothly approximate the full probability distribution.

As discussed in section 11.1, the use of emulators—statistical models to interpolate model outputs across parameter space—provides one means for estimating model sensitivity when models are computationally expensive. The exact same approach can also be used with Monte Carlo uncertainty propagation and uncertainty analysis, where the emulator is used in place of the full model as an alternative means of approximating the model (the Taylor Series approach earlier was likewise a means of approximating the model). This has the advantages of approximating a full distribution and accounting for nonlinearities (Jensen's inequality), but does introduce an additional approximation error, so has little advantage for models that are not computationally demanding.

### 11.2.5 Ensemble (Stats)

Ensemble simulation is closely related to Monte Carlo simulation as a means for uncertainty propagation. Indeed, in many contexts the two are considered synonymous. The distinction I make here is that ensemble simulations typically employ a considerably smaller sample size, and are thus more focused on estimating the sample mean and variance than on approximating the shape of the full distribution. Ensemble sizes tend to be in the range of 10 to 100 and, like emulated MC, are frequently employed in cases where model simulations are computationally expensive. The trade-off in the ensemble approach is that, like with the Taylor Series approach, you are losing accuracy and making strong assumptions about the shape of the probability distribution (which is typically assumed to be Normal).

In most cases ensemble forecasts are random samples of the inputs, as in Monte Carlo simulations, but there are a few notable exceptions. One common example would be the "ensemble of opportunity" that frequently occurs when making predictions with multiple models. The models in a multimodel ensemble are never a random sample of all theoretically possible model structures. Indeed, they are rarely even a random sample of the model structures currently employed by a research community. Furthermore, in many cases the models are not even independent, as modeling teams frequently learn and borrow from one another.

In other cases a nonrandom sample may be employed strategically to maximize information about forecast uncertainty with the fewest possible number of model runs. For example, in a univariate model one might try to get away with only three runs, such as the mean and some $\pm$ perturbation, and then assume that, in the output, the perturbations continue to represent the same quantiles (which is only valid if the function is monotonic). In multivariate models one might rely on some form of dimension reduction, such as PCA, to account for input correlations and then sample within a transformed parameter space. Finally, there are other clever approaches, such as the Unscented Transform, which uses $2n + 1$ weighted points in parameter space, known as *sigma points*, to provide a unique solution for the coefficients of a multivariate mean vector and covariance matrix in $n$ dimensions (Julier et al. 2000). These points form the basis for an ensemble of model runs, and the model outputs are used to algebraically solve for a weighted mean and covariance, as opposed to statistically estimating the unweighted mean and variance of a random sample (figure 11.7).

## 11.3 UNCERTAINTY ANALYSIS

The goal of uncertainty analysis is to attribute the uncertainty in some overall response variable, $Y$, to the different input $X$'s. Uncertainty analyses serve an important role in diagnosing models and forecasts. In particular they provide critical direction in efforts to reduce uncertainties by focusing data collection and synthesis efforts on the parameters and processes that dominate a model's uncertainty (section 11.4). These analyses are central to the idea of the model-data feedback loop (section 1.3, figure 1.2)—using the formal quantification and analysis of model uncertainties to target those measurements that most efficiently reduce model uncertainties.

Formally, uncertainty analysis combines information about model sensitivity (section 11.1) and input uncertainty (chapter 6). If we look at the Taylor Series derivation for uncertainty propagation, the total variance was the sum of all the component

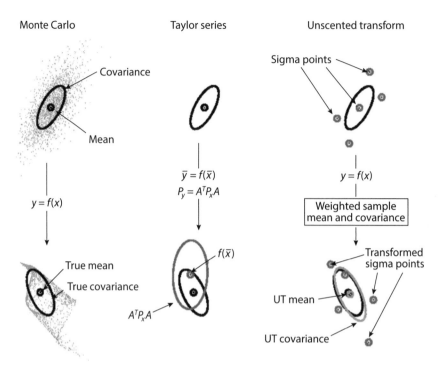

Monte Carlo            Taylor series            Unscented transform

FIGURE 11.7. Comparison of Monte Carlo, Taylor Series, and Unscented Transform approaches to uncertainty propagation. The top row illustrates the same input, $x$, (black ellipse), while the bottom row contrasts the projection of $x$ through $f(x)$ using each technique. The black ellipse in the bottom row represents the correct mean and covariance and is compared to each approximation. In this example the Unscented Transform provides a close approximation and performs much better than the Taylor Series. Figure modified from www.cslu.ogi.edu/nsel/ukf/node6.html.

variances and covariances multiplied by the square of the component sensitivities (equation 11.5). This demonstrates that both components contribute equally to output uncertainty. In other words, a parameter can be important either because it is sensitive, or because it is unknown (figure 11.8). Likewise, sensitive parameters can be unimportant if they are well constrained, and highly uncertain parameters can be unimportant if they are insensitive.

How one conducts an uncertainty analysis is very much determined by how the uncertainty propagation and sensitivity analyses were performed. In the Analytical Moments and Taylor Series approaches, *each component in the summation determines the contribution of that parameter, or multi-parameter interaction, to the overall uncertainty*.

If you used a numerical approximation to sensitivity, such as an OAT sensitivity analysis, rather than an analytical analysis, then the concept embodied in the Taylor Series approach can also be generalized to

$$Var[f(X)] \approx \sum_{i=1}^{n} Var[p(y_i)] \tag{11.14}$$

where $p(y_i)$ is the transform of $p(x_i)$, the PDF of the $i^{th}$ input, through $g_i(x_i)$, the univariate response of the model to the $i^{th}$ input conditioned on all other variables,

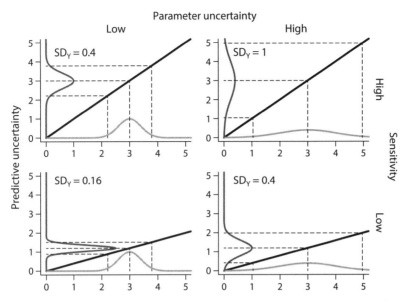

FIGURE 11.8. Predictive uncertainty (distribution on the y-axis and reported standard deviation, $SD_Y$) is controlled by parameter uncertainty (distribution on the x-axis, varies by column) and model sensitivity (slope of diagonal line, varies by row).

$f(x_i|X_{-i})$ (LeBauer et al. 2013). Less formally, $g_i$ is the OAT sensitivity and will frequently be approximated by a univariate emulator fit to the OAT sensitivity analysis (for example, a smooth spline). The univariate transform of $p(x_i)$ through $g_i(x_i)$ will often be approximated using a Monte Carlo approach. Therefore, the variance within the sum is simply the variance of the OAT transform, and thus is the component variance of the $i^{th}$ input. With this approach it is likewise very straightforward to attribute the component uncertainties to their sensitivities and their parameter uncertainties (figure 11.9).

When using Monte Carlo and Ensemble methods for uncertainty propagation, the method for partitioning uncertainty is identical to those used in the Monte Carlo sensitivity analysis ($R^2$ from ANOVA, regression). As noted in section 11.1, the only real distinction is in the distribution of $X$ that is sampled over. If the uncertainty in $X$ represents your best estimate generated from the available data, then the resulting analysis is an uncertainty analysis. If, on the other hand, the uncertainty in $X$ is prescribed somewhat arbitrarily, such as the classic $\pm 10\%$, then the output of the exact same analysis should be considered a sensitivity analysis.

An alternative to using linear models to partition Monte Carlo uncertainties is the use of Sobol indices (Cariboni et al. 2007; Saltelli et al. 2008). Sobol's approach allows the decomposition of variance into arbitrary parameter groupings, but most analyses focus on two components, the univariate effects of individual parameters, $S_i$, and the *total effect* of individual parameters accounting for all their interactions, $ST_i$. These are calculated for the $i^{th}$ parameter as

$$S_i = \frac{Var[E[Y|x_i]]}{Var[Y]} \qquad (11.15)$$

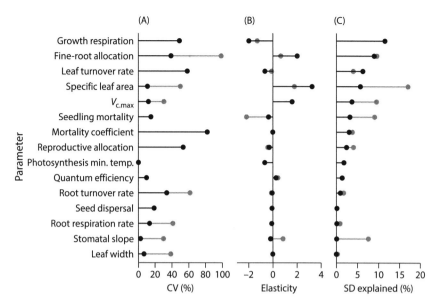

FIGURE 11.9. Partitioning of the uncertainty in predictions of switchgrass (*Panicum virgatum*) aboveground biomass by the Ecosystem Demography model under both prior (gray) and posterior (black) parameter estimates generated from a meta-analysis of trait data (chapter 9). Parameters are ranked by their component uncertainties (column C), which are calculated from the parameter uncertainties (visualized as the CV, column A), and the OAT model sensitivity to that parameter (visualized as parameter elasticity, column B). Reproduced from LeBauer et al. (2013).

$$ST_i = \frac{E[Var[Y|x_{-i}]]}{Var[Y]} \tag{11.16}$$

where $x_{-i}$ refers to all variables except $x_i$. A number of different numerical approaches are available to approximate these based on Monte Carlo output (Pujol and Iooss 2014). It is worth noting that in all of the numerical approaches described earlier, the individual parameter effects do not sum to 100% because they ignore parameter interactions while the total effect will sum to >100% because it double counts all parameter interaction (that is, the interaction of parameter $i$ and parameter $j$ is in both $ST_i$ and $ST_j$).

## 11.4 TOOLS FOR MODEL-DATA FEEDBACKS

The central goal of uncertainty analysis is to identify which parameters or inputs are responsible for model uncertainty. Given this information, *we can then target additional measurements and data synthesis around the most important processes.* However, this raises the question of how best to do this. Such questions are not new to ecologists, but quantitative answers to such questions are relatively uncommon (Schimel 1995). In particular, how much do we need to measure different processes when faced with limited resources? Do we need to make measurements intensively at a single site or extensively across a network? How do we balance the need to constrain different processes? Do we always tackle the most uncertain process first?

How does the amount of data already collected for each process, as well as the cost of different measurements and the availability of existing equipment and infrastructure, affect these choices? These are the types of practical experimental design questions that are the focus of this section.

### 11.4.1 Power Analysis

The traditional goal of power analysis has been to estimate the sample size required to detect an effect of a given size. Power analyses are a key part of experimental design, ensuring that adequate sampling is done to test alternative hypotheses or select among competing models, but such analyses are also critical to reducing forecast uncertainty. At the heart of such an analysis is the estimation of how the uncertainty in a parameter decreases as a function of sample size (figure 11.10, top left), because it is knowing when that uncertainty will drop below some threshold that determines the sample size required. In the simple case of Gaussian means, it is well established that the standard error goes down as

$$SE(n) = \frac{\sigma}{\sqrt{n}} \tag{11.17}$$

where $\sigma$ is the standard deviation and $n$ is the total sample size. If we have some initial pilot data to estimate $\sigma$, we can use this equation to estimate the total sample

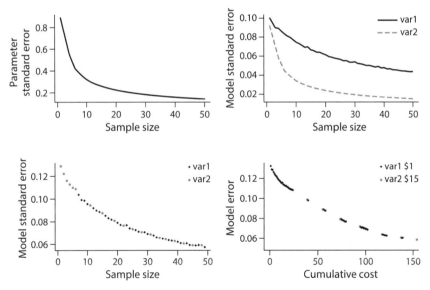

FIGURE 11.10. Observational design. Top left: Power analysis estimating reduction in parameter uncertainty as a function of sample size. Top right: Translation of parameter uncertainty into model output uncertainty for two different variables. Bottom left: Combined output uncertainty as a function of sample size with the optimal design indicated by color and shading. For example, at a sample size of 8 the optimal design is six samples of variable 2 and two of variable 1. Bottom right: Output uncertainty as a function of overall project cost assuming different marginal costs for the two data types. Compared to the equal-cost design (bottom left), there is considerably greater sampling of variable 1.

size required to get the SE down to some specific threshold. The threshold we're trying to meet might be that required for hypotheses testing (that is, to detect that a specific effect size is different from zero) or to meet some regulatory goal. More generally, if we are approaching the problem iteratively, where we already have some number of initial samples in hand, $n_{init}$, and we want to determine how many additional samples we need, $n$, then this is approximately

$$SE(n) = \frac{\sigma_{init}}{\sqrt{n + n_{init}}} \qquad (11.18)$$

With more complex models we are frequently interested in the uncertainty surrounding quantities, such as model parameters, that are more complex to calculate than simple sample errors (equation 11.17). Our simplest estimator for the reduction in uncertainty for an arbitrary parameter $\theta$ is just the generalization of the standard error

$$Var(\theta|n) = \frac{Var[\theta_{init}]n_{init}}{n + n_{init}}$$

where $Var[\theta_{init}]$ and $n_{init}$ are our initial estimate of the variance and sample size, and $n$ is the additional number of samples added. However, this is a pretty coarse estimate and is unable to accommodate a lot of the questions we raised initially about the distribution of sampling.

A more general approach to power analysis is Monte Carlo simulation based on fitting the model to pseudo-observations according to different proposed sampling schemes or sample sizes. The general algorithm, which is essentially a variation on the common *bootstrap* approach, is

for(k in 1:M)

    Draw random pseudo-data of size N
    Fit model to data
    Save parameters of interest

The preceding approach may be embedded in outer loops that vary the sample size, N, the sampling design, or other assumptions. A key difference in this approach, which sets it apart from a traditional bootstrap, is that we are explicitly interested in drawing a sample of data larger than our original pool of observations. As with the bootstrap there are two alternative ways of generating pseudo-data. In the nonparametric approach, existing data are resampled with replacement, meaning that a previously sampled observation can be resampled any number of times. Alternatively, in the parametric approach pseudo-data are simulated from the current model, including any "residual" error as well as all sources of uncertainty and variability that were in the original model. This process is repeated $M$ times to generate a distribution of parameter estimates under different pseudo-data realizations. If the model is fit to pseudo-data via Maximum Likelihood or some other cost function optimization, then $M$ must be large to generate a distribution of estimates—recall that it is the variance in those estimates that is our primary interest in the current context. If the model is fit in a Bayesian context, then each fit already produces a posterior distribution (or, more commonly, MCMC samples from that distribution), and thus $M$ can be much smaller, but should still be >1 to ensure that parameter estimates from different pseudo-data realizations are consistent.

### 11.4.2 Observational Design

If we are working with a simple model, and in particular if our model is informed by only one type of observation, then the preceding approach is often sufficient to determine how uncertainty changes with sample size. From that point we might estimate: How large a sample size would we need to bring the uncertainty under some threshold? How much uncertainty reduction might we expect from the sample size we can afford? How would alternative sampling designs with the same sample size perform? However, for more complex models our uncertainty analysis (section 11.3) will have identified the key parameters of interest and we need to decide among alternative measurements to constrain different processes.

All else being equal we may want to tackle the largest uncertainties first, but all else is rarely equal. If we compute power analyses for different parameters, we will often find that parameter uncertainty reduces at very different rates. Indeed, recalling that uncertainty tends to reduce as $1/\sqrt{n}$, in general *we expect parameters that are important due to high parameter uncertainty to improve more rapidly with additional sampling than parameters that already have some constraint but remain important due to their sensitivity.* For example, if we want to halve the uncertainty in a parameter, we need to increase our sample size by a factor of four. For a poorly constrained parameter that may mean going from a sample size of 10 to 40, while a well-constrained parameter may need to go from a sample size of 1000 to 4000.

To compare the uncertainty reduction among multiple parameters, what we really want to do is to *evaluate how we reduce uncertainty in model predictions*, not model parameters. Much as we saw when comparing parameters in the uncertainty analysis, this allows us to make an "apples to apples" comparison of the impact of improving different parameters. To do so requires using the uncertainty propagation and uncertainty analysis techniques discussed previously to transform our power analysis relationship between sample size and parameter uncertainty into the *relationship between sample size and variance contribution to overall model uncertainty* (figure 11.10, top right).

Given these relationships for multiple parameters or inputs we can then apply economic principles to prioritize sampling. In the simplest case, where all measurements cost the same, this is as simple as starting with the parameter that gives the greatest marginal reduction in uncertainty for one additional measurement (that is, the greatest change in model output uncertainty when going from the current sample size, $n_0$, to $n_0 + 1$). This process is then repeated iteratively, generating a plot of the overall expected reduction in model uncertainty with sample size and the number of observations of each parameter for a given sample size (figure 11.10, bottom left). As each parameter is sampled there is a diminishing return to the reduction in uncertainty for additional samples, so it will be quite common for this approach to suggest a diverse portfolio of measurements rather than heavily investing in any one data type.

It is also fairly straightforward to relax the assumption that all observations incur the same cost. If we can assign a marginal cost to making measurements of each data type (which may be monetary, person-hours, or any other units that make sense for a particular problem), then we again use economic principles to choose the next observation so that it gives us the greatest marginal reduction in uncertainty per unit cost (figure 11.10, bottom right). Importantly, such an analysis can still be performed sequentially, choosing the best option one datum at a time. However, there are also fixed costs that often need to be considered. These may be the costs of purchasing a specific piece of field or lab equipment, setting up at a new field site, and so on. Indeed, it is very often the case with scientific equipment that the fixed costs of purchasing equipment are very high and the marginal costs of each individual measurement are fairly low. Solving this problem requires a more global search through the different options, as sampling strategies may change radically when resource thresholds are crossed that allow the purchase of equipment that was not previously justified. It also illustrates that the optimal sampling design can change considerably based on the infrastructure and equipment already available to an individual, organization, or research network. A range of approaches are available for this type of constrained optimization, with the most common based on the method of Lagrange multipliers. Care must be taken when applying these approaches, however, to ensure that one works with integer-value choices of sample size and fixed costs—most of us cannot buy a third of an instrument or make half of a measurement.

These types of analyses fall within a broader class of research called Observing System Simulation Experiments (OSSEs) (Charney et al. 1969; Masutani et al. 2010a, 2010b). The general pattern of such model-based experiments is to simulate a hypothetical "true" state of some system and to generate pseudo-data observations of that system. Those observations are then assimilated into models to assess impacts on estimates and inference. OSSEs are frequently used in other disciplines, such as weather forecasting, remote sensing, oceanography, and atmospheric $CO_2$ inversions, to assess the value of new measurement campaigns or technologies, and for augmenting existing research networks (Mishra and Krishnamurti 2011).

Overall, improved observational and experimental design, through uncertainty analysis, power analysis, OSSEs, and basic economic principles, represents an opportunity for the ecological community to improve both the relevance and efficiency of research. Beyond an individual team or project, analyses such as these also provide important feedback to the community for setting research priorities that are quantitative and defensible. Often, but not unsurprising, they will tell us that the priorities are not the processes we are best at measuring or the systems we've spent the most time working in (Leakey et al. 2012; Dietze et al. 2014; Schimel et al. 2014). More constructively, they will also tell us that the alternative is not simply to measure everything, everywhere, but to identify what's needed most and where it's most needed.

## ▨▨▨ 11.5 KEY CONCEPTS

1. Sensitivity analyses assess how a change in $X$ translates into change in $Y$.
2. Curse of dimensionality: parameter space increases in size exponentially with the number of parameters, making it harder to explore parameter space.
3. The range of values explored in a sensitivity analysis should be tied explicitly to the probability distribution of the inputs.
4. How you treat uncertainties changes the nature and meaning of many analyses.
5. Uncertainty propagation refers to the process of translating uncertainty in our model's inputs, $X$, into uncertainty in the model's outputs, $Y$. All approaches depend upon knowing the uncertainty in $X$ and the sensitivity of $Y$ to $X$.
6. The uncertainty in a prediction is the product of the uncertainty in the inputs and the model's sensitivity to those inputs.

7. The uncertainty in a deterministic forecast with known parameters and boundary conditions is dependent upon three components: the uncertainty about the state, $Var[x_t]$, the sensitivity of the system, $f'(\overline{x}_t)^2$, and the process error, $q$. In ecological systems with stabilizing feedbacks, $f'(\overline{x}_t) < 0$, process error dominates forecasts.

8. The goal of uncertainty analysis is to attribute the uncertainty in some overall response variable, $Y$, to the different input $X$'s.

9. A parameter can be important either because it is sensitive, or because it is unknown.

10. In sensitivity and uncertainty analyses and in uncertainty propagation, there are trade-offs among methods. In general more precise answers require more computation, while approximations require stronger assumptions.

11. Power analyses estimate how the uncertainty in the parameter of interest decreases as a function of sample size.

12. Observing System Simulation Experiments are used to explore alternative experimental or observational designs, or to optimize sampling accounting for fixed and marginal costs of different measurements.

## 11.6 HANDS-ON ACTIVITIES

See the end of chapter 12 for a hands-on activity applying the concepts of sensitivity analysis, uncertainty propagation, and uncertainty analysis to simple ecosystem models.

## APPENDIX A: PROPERTIES OF MEANS AND VARIANCES

In the following tables (tables 11.2 and 11.3), take $X$ and $Y$ to be random variables, $a$, $b$, and $c$ to be constants, and $g(x)$ be a function. Furthermore, let E[X] indicate the expected value of $X$ (that is, the mean of $X$) and Var[X] indicate the variance of $X$.

TABLE 11.2. Properties of Means

| | |
|---|---|
| $E[c] = c$ | Mean of a constant. |
| $E[X + c] = E[X] + c$ | Mean plus a constant. |
| $E[cX] = c \cdot E[X]$ | Mean times a constant. |
| $E[X + Y] = E[X] + E[Y]$ | Means are additive. |
| $E[a + bX + cY] = a + bE[X] + cE[y]$ | Application of preceding rules. |
| $E[E[X|Y]] = E[X]$ | Iterated expectation. |
| $E[g(X)] \neq g(E[X])$ | Jensen's inequality, weak form (unless $g(x)$ is linear). |
| $E[XY] = E[X]E[Y] + Cov[X, Y]$ | Multiplication. |

TABLE 11.3. Properties of Variances

| | |
|---|---|
| $Var[c] = 0$ | Variance of a constant |
| $Var[X + c] = Var[X]$ | Variance plus a constant |
| $Var[cX] = c^2Var[X]$ | Variance times a constant |
| $Var[X + Y] = Var[X] + Var[Y] + 2Cov[X, Y]$ | Sum of variances |
| $Var[a + bX + cY] = b^2Var[X] + c^2Var[Y] + 2bcCov[X, Y]$ | Application of preceding rules |
| $Var[\sum a_i X_i] = \sum\sum a_i a_j Cov[X_i, X_j]$ | Generalizes sum of variances |
| $Var[X] = Var[E[X|Y]] + E[Var[X|Y]]$ | Variance decomposition |

## APPENDIX B: COMMON VARIANCE APPROXIMATIONS

All of the following approximations (table 11.4) are derived based on the first-order Taylor Series approximation described in section 11.2.3.

TABLE 11.4. Variance Approximations

| | |
|---|---|
| $Var[XY] \approx (XY)^2\left(\dfrac{Var[X]}{X^2} + \dfrac{Var[Y]}{Y^2} + 2\dfrac{Cov[X,Y]}{XY}\right)$ | Multiplication |
| $Var[\prod X_i] \approx (\prod X_i)^2\left(\sum\sum\dfrac{Cov[X_i,X_j]}{X_i X_j}\right)$ | Product |
| $Var\left[\dfrac{X}{Y}\right] \approx \left(\dfrac{X}{Y}\right)^2\left(\dfrac{Var[X]}{X^2} + \dfrac{Var[Y]}{Y^2} + 2\dfrac{Cov[X,Y]}{XY}\right)$ | Division |
| $Var[aX^b] \approx (abX^{b-1})^2Var[X]$ | Powers |
| $Var[ae^{bX}] \approx (abe^{bX})^2Var[X]$ | Exponential, base $e$ |
| $Var[ac^{bX}] \approx (ab\ln(c)c^{bX})^2Var[X]$ | Exponential, base $c$ |
| $Var[a\ln(bX)] \approx \left(\dfrac{a}{X}\right)^2Var[X]$ | Logarithm |

# 12

## Case Study: Carbon Cycle

SYNOPSIS: *The feedbacks between the biotic carbon cycle and climate change are a complex, global, and centennial-scale problem. The terrestrial carbon cycle forms one of the greatest sources of uncertainty in climate projections. Because of this terrestrial biosphere models present some of the greatest challenges and opportunities in ecological forecasting. In this case study we highlight the important role of uncertainty analysis and propagation in these complex, nonlinear models.*

### 12.1 CARBON CYCLE UNCERTAINTIES

If there's one ecological forecast that keeps me up at night, it's the projections of the terrestrial carbon sink. Figure 12.1, which combines the results from two recent iterations of the Coupled Model Inter-comparison Project (CMIP), sums up the state of carbon cycle research and the challenges we face in becoming a more predictive science (Friedlingstein et al. 2006, 2014). First, both panels demonstrate a tremendous amount of uncertainty—we currently can't agree on the sign of the flux, let alone the magnitude. The highest projection suggests the land will be a sink of comparable magnitude to current anthropogenic emissions (problem solved!), while the lowest suggests it will be a source of similar magnitude (problem doubled!). It's not hard to argue that this is a policy-relevant amount of uncertainty. Second, not only do the current projections have large uncertainties, but if we compare these two graphs, it also suggests that our ability to forecast is not improving quickly enough.

Not only is the uncertainty in the land sink at the centennial scale policy relevant, the land sink uncertainty is also relevant at the interannual to decadal scale. If we look at the uncertainties in the US Climate Action Plan (CAP; figure 12.2), the uncertainty associated with the terrestrial carbon cycle falls under Land Use, Land Use Change, and Forestry (LULUCF). This uncertainty is approximately equal to all other sources of uncertainty in the CAP (policy, economics, technology, and so on) put together. This uncertainty isn't just a US issue. Currently LULUCF is one of the international reporting requirements under the UN Framework Convention of Climate Change, but the accounting process is closer to the deterministic spreadsheets that were the state of fisheries management in the early 1990s (chapter 10) than to an exercise in data assimilation and forecasting.

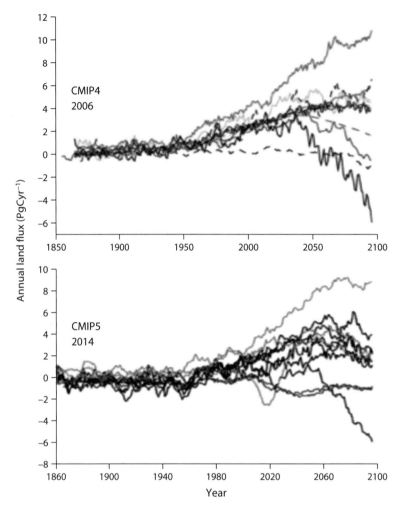

FIGURE 12.1. Ensemble terrestrial carbon flux projections. Each line represents a different Earth System model. Positive fluxes are sinks. Reproduced from Friedlingstein et al. (2006; 2014).

## 12.2 STATE OF THE SCIENCE

Given these uncertainties, an obvious question is "what is the current state of ecological forecasting surrounding the carbon cycle, and how can we improve it?"

If we think mathematically about the models used by ecosystem ecologists and biogeochemists, at a high level they are not too different from the age- and stage-structured matrix models used in population ecology (chapter 7) and natural resources (chapter 10). Most carbon cycle models are based on the concept of pools and fluxes, which can be conceptualized as matrix models, with the pools being the different states or stages and the transitions between pools forming the fluxes (Luo et al. 2014). From this perspective, the biggest difference between population and ecosystem models is the units—in population and community ecology the units are almost always *individuals*, while in ecosystem ecology pools and fluxes are on a *mass*

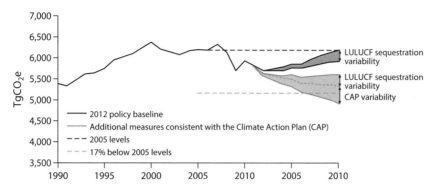

FIGURE 12.2. US carbon emissions under a business-as-usual scenario compared to the Climate Action Plan (CAP).

or *energy* basis, which can come in a wide array of different measurement units. Indeed, one of the things I struggled with when I first branched out into ecosystem ecology was the wide range of different units used ($kg/m^2$, MgC/ha, $\mu mol/m^2/s$, and so on) and all the unit conversions between them!

The other big difference in ecosystem models is the *wide range of timescales* that are simulated explicitly—most population models operate on an annual or a generational basis, but even the coarsest ecosystem model represents the seasonal cycle explicitly, and most models now capture subdaily fluctuations. So why are ecosystem models being run at finer and finer temporal resolutions, if the goal is to make centennial-scale projections? In a nutshell, it's because of *nonlinearities*. While the demographic transitions in population matrices are often constants or based on statistical models (chapters 7 and 10), the fluxes in ecosystem models are typically *semimechanistic functions* based on ecophysiological first principles. Almost every flux is scaled by a temperature function that's in the exponential family (Q10, Arrhenius, Van 't Hoff, Lloyd-Taylor, and so on), and many fluxes are modified by mechanistic or statistical functions of moisture. Many models also impose stoichiometric constraints on the nutrient ratios of many pools and fluxes, which often generate thresholds and breakpoints in flux equations.

To illustrate the impacts of timescales and nonlinearities consider photosynthesis, the driving flux of the terrestrial carbon cycle, which is a nonlinear function of light, temperature, and $CO_2$ (Farquhar et al. 1980). In addition to photosynthesis itself being a nonlinear process, photosynthesis models further need to be solved simultaneously with a model of stomatal conductance, which typically combines a semimechanistic function of photosynthesis, humidity, and $CO_2$ with additional ad hoc terms for soil moisture and nutrient limitation (Medlyn et al. 2011). To further complicate things, the temperature used when modeling photosynthesis is leaf temperature, not air temperature, and estimating leaf temperature requires solving the balance of all incoming and outgoing energy fluxes (solar radiation, thermal infrared radiation, sensible heat, and the latent heat flux from transpiration). The last of these is also a function of stomatal conductance, which was a nonlinear function of photosynthesis, which itself was a nonlinear function of leaf temperature. Therefore, the leaf energy budget also needs to be solved simultaneously with the nonlinear models of photosynthesis and stomatal conductance. To make life further complicated, all of this varies with your position in the canopy, as do all the parameters to the equations, and all of the meteorological inputs change over the course of a single day.

Because of Jensen's inequality, we know we can't just stick the means of the inputs (for example, over a diurnal cycle) and parameters into these models and expect to produce the correct outputs (for example, mean photosynthetic rate). So while population and ecosystem models look similar superficially, ecosystem models quickly spiral into layers of complexity and interdependence. While the simplest ecosystem models may involve only a score of equations and a page of code (Fox et al. 2009), the most complex models involve hundreds of thousands of lines of computer code (Medvigy et al. 2009; Oleson et al. 2013).

Considered as a whole, ecosystem models benefit greatly from being able to take advantage of first principles (for example, conservation of mass and energy), but suffer from a burden of complexity, both conceptually and computationally. Unfortunately, both of these factors (first principles and complexity) contribute to the fact that most carbon cycle forecasts continue to be made deterministically rather than probabilistically. For example, if you look the model runs in figure 12.1, each model is represented by a single line and the only uncertainty in the ensemble is the across-model structural uncertainty. Some of these lines represent an average over different realizations of the *atmosphere*, but within the land models there is no accounting for parameter error, uncertainties in initial conditions (pools, land use) and lower boundary condition (soils, topography), or process error (model inadequacy, variability and heterogeneity in processes, stochastic events, and so on). In other words, the enormous amount of uncertainty in figure 12.1 is an *underestimate*, and we have no idea how much more uncertain we actually are. Part of the reason for this lack of uncertainty accounting is computational, though in many cases the *other uncertainties simply have not been quantified*.

Thankfully, there has been strong growth in the application of Bayesian calibration approaches in the carbon cycle literature (where it is often referred to as model-data fusion or data assimilation), with a number of reviews synthesizing this rapidly growing literature (Luo et al. 2009; Williams et al. 2009; Keenan et al. 2011; Zobitz et al. 2011; Dietze et al. 2013; Niu et al. 2014). The large majority of this literature is focused on estimating parameters and parameter uncertainties (with a subset of these also considering initial condition uncertainty), which is not surprising because these large models may contain hundreds of parameters, most of which are empirical coefficients rather than physical constants, and many of which are underconstrained by direct measurements. Also not surprising, given the computational demand of these models, a lot of the work to date has focused on simpler ecosystem models and at the scale of individual sites (for example, single experiments or individual eddy-covariance towers). Carbon cycle scientists, due to their closer working relationships with atmospheric scientists and hydrologists, have tended to borrow data assimilation tools and techniques from these disciplines (for example, various flavors of Kalman filters; see chapters 13 and 14). While this approach has greatly accelerated the adoption of data assimilation, unfortunately this borrowing has at times not been sufficiently critical of where and how the ecological forecasting problem differs from forecasting physical systems (for example, weather, see section 2.5). Here I highlight three key areas for improvement in carbon cycle forecasting.

One area for improvement is to more completely *embrace the complexities of data models and process errors in ecological systems* (chapter 6). Initial efforts at ecosystem model-data fusion were over-reliant on weak priors and Gaussian likelihoods. In addition, not only was there a tendency to treat error as a single quantity (that is, residual), rather than partitioning variability, but it has also been common to treat uncertainties as known quantities (for example, a reported error in a data product). As discussed elsewhere (chapters 6 and 14), having strong priors on observation errors in data models *is* extremely helpful in partitioning sources of uncertainty, but that is different from assuming all other sources are zero. Furthermore, not partitioning sources of variability has led to apparent inconsistencies, such as model calibrations at different sites, or for different years within a site, yielding different parameter estimates. These "inconsistencies" are likely a reflection of true heterogeneities and historical legacies within and among ecosystems that are causing systematic variability in model processes across space and time. Accounting for such variability will not only lead to better predictions but may also open the door to a better understanding of ecosystem processes. Hierarchical approaches to evaluating carbon cycle data (Matthes et al. 2014) and calibrating carbon cycle models (Naithani et al. 2015) are just beginning to emerge using simple models, but I anticipate this being an important research direction in the future, in particular as model-data fusion is applied across larger spatial scales and a wider range of data types.

A second major data assimilation challenge facing carbon cycle research is in *fusing multiple sources of data* (chapter 9). There is an incredibly wide range of *relevant* data on the carbon cycle, spanning a broad range of spatiotemporal scales and processes, from <1 second to millennia and from the molecular scale up to global remote sensing. However, even at the scale of a single site, initial attempts at fusing multiple data sources were largely ad hoc (Medvigy et al. 2009). That said, the semi-mechanistic nature of carbon cycle models really highlights the potential pitfalls in fusing data. In particular, a common occurrence in carbon cycle science is to have a mix of automated, high-volume data and manually collected low-volume data. If handled naïvely, the sample size of the automated data dwarfs the manual data and the fits largely reflect that of the automated data alone. Not only does this mean we are not learning much from labor-intensive hard-won data, but often models will fail to fit this data well because the fit is driven by the automated data. Furthermore, due to the large sample size of the automated data, confidence intervals can be very tight around values that are incompatible with observations. As noted in chapters 6 and 9, this can arise from failing to account for the complexities of real data, such as autocorrelation. Two very common challenges with automated data are the presence of *systematic errors* and a *lack of replication* (for example, $n = 1$ flux towers).

The lack of replication means that sampling errors need to be accommodated by adding bias terms to the data model (that is, this one tower is different from the unobserved sample mean of towers by some unknown amount). Furthermore, due to a lack of replication it can be difficult to ascertain if these biases are constant, random, autocorrelated (for example, sensor drift), driven by unknown covariates, or some combination of all of these. Systematic errors can similarly enter due to sensor bias (miscalibration), sensor drift, or as functions of other covariates.

Recall the example from chapter 9 of a single air temperature sensor in a forest understory. If you deployed hundreds of sensors, you could estimate the sample mean precisely, but if only one sensor is deployed, then you don't know how different that one sensor is from the true mean (lack of replication). The sensor will also have calibration uncertainty (errors in variables) and may be prone to drift (systematic error).

Furthermore, it may have systematically biases (for example, whenever struck by a sunfleck). As noted in chapter 9, the problem with both systematic errors and sampling biases is that they *don't average out over time.*

In addition to the systematic errors in the data, assimilating multiple sources of data can highlight *systematic errors in the model*. If a model was structurally perfect, then assimilating one type of data would asymptotically estimate all model parameters correctly, and the model would be able to correctly predict novel outputs. If the model is structurally incorrect, then, as noted earlier, the model will asymptotically converge upon a solution that matches the high-volume data type well but is overconfidently biased for other outputs. Importantly, the standard parameterization of model process error is focused on random model errors, not systematic errors, so adding random process error to a model doesn't necessarily solve this problem. All models are approximations and thus, in a big data era, all models will be overconfidently biased unless there is an explicit inclusion of both random and systematic process errors.

The third major challenge is the *presence of long memory* in certain parts of the carbon cycle, such as soil carbon. This challenge manifests itself during initialization and forecasting. A common initialization practice in carbon cycle models is to spin up a model to steady-state, which can require model runs that are millennia long. During data assimilation or ensemble forecasting every different ensemble member would need to be spun up, which can be enormously computationally expensive, even if efficient semianalytical approximations are employed (Xia et al. 2012). Furthermore, since models will inevitably contain structural errors, and the real-world contains many chance events, the steady-state that a model is spun up to is unlikely to match the true conditions. During forecasting, the relative impact of "fast" carbon pools decays over time, while the impact of "slow" pools becomes more and more important. Therefore, even when using data assimilation to constrain model parameters and initial states, the impact of the data on the forecast decays over time, while the impact of model structure progressively increases (Weng and Luo 2011) (figure 12.3). All in all, this implies that it is precisely the parts of models that are hardest to evaluate using modern data that are most important for long-range projections. This problem cannot be solved simply by larger data; it requires using experiments and constraints at other scales (both larger and smaller) to directly target improvements to our process-level understanding.

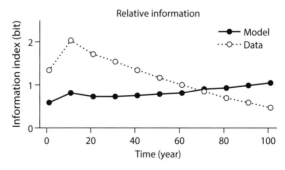

FIGURE 12.3. Information contribution of data versus model to long-term forecast of total ecosystem carbon for Duke Forest, North Carolina (Weng & Luo 2011).

## ▨ 12.3 CASE STUDY: MODEL-DATA FEEDBACKS

For all the challenges facing carbon cycle forecasting, there are a great number of opportunities. One noteworthy example is a growing recognition of the potential of model-data fusion to inform field research, by refining our understanding of which processes are limiting our ability to forecast and what data types are most useful. At a high level the US Department of Energy's Terrestrial Ecosystems program formalized Mod-Ex (Model-informed Experiments) as a guiding principle in its research program. In addition to identifying the drivers of long-term forecast uncertainty (figure 12.3), model-data fusion has been used to "rate my data" by identifying the information contribution of different data types to forecast uncertainty (Keenan et al. 2013). Model-data fusion is increasingly performed as part of larger workflows involving sensitivity and uncertainty analyses (Wang et al. 2012; LeBauer et al. 2013; Safta et al. 2014). Such uncertainty analyses have been used at continental scales to identify common sources of uncertainty across multiple biomes and vegetation types (Dietze et al. 2014) and to reduce the dimensionality of subsequent data assimilation (Wang et al. 2012; Safta et al. 2014).

In my own lab we are developing software, the Predictive Ecosystem Analyzer (PEcAn), that automates these types of workflows and makes many of the tools and techniques discussed in this book accessible to the broader research community rather than just ecosystem modelers (Dietze et al. 2013). PEcAn presents the end user with an intuitive web-based interface that allows users to run alternative models for different sites and then visualize outputs. Furthermore, the PEcAn system allows users to manage the flows of information into and out of models (chapter 3); track the provenance of model runs and analyses (chapter 4); calibrate model parameters using data (chapters 5 and 6); synthesize multiple data sets (chapter 9); perform ensemble, sensitivity, and uncertainty analyses (chapter 11); assimilate data (chapters 13 and 14); and assess model performance (chapter 16).

While PEcAn addresses a wide range of informatics, computational, and statistical issues, facilitating the feedback between the collection and assimilation of plant trait data is a core workflow of the system (LeBauer et al. 2013). This workflow begins with a plant functional type (PFT), which is defined dynamically within the PEcAn system database by a list of species, a model-specific list of parameters, and a list of prior probability distributions for each parameter. The system then queries its trait database by species and parameter and uses a Hierarchical Bayes meta-analysis (chapter 9) to synthesize trait data based on both raw observations and summary statistics culled from the literature. The aim of this analysis is to update the prior distributions, based on the combined information from multiple studies, in a way that captures the uncertainties in the data and the trait variability within and across sites. The meta-analysis accounts for the differences in sample size and variability across data sets, and the high frequency of missing data. The meta-analysis also controls for the biases associated with both greenhouse and potted plant studies and experimental manipulations, acknowledging that such studies do provide useful information about plant traits, but that this information cannot be treated the same as field observations.

The outputs of the PEcAn meta-analysis are frequently used as inputs to ensemble, sensitivity, and uncertainty analyses (chapter 11). Ensemble analyses propagate trait uncertainty into model forecasts, allowing us to put a confidence interval on a model projection. Sensitivity analyses explore how changes in individual trait values affect model predictions. Uncertainty analyses combine our information about the uncertainty in traits, which comes out of the meta-analysis, and our information about model sensitivity to estimate the contribution of individual traits to the overall predictive uncertainty captured by the ensemble analysis. These analyses have identified trait uncertainty and model sensitivity as equal partners in determining overall model uncertainty—each in isolation paints an incomplete and potentially misleading picture, as traits can be important either by being particularly sensitive or uncertain. We have identified a number of cases where highly sensitive parameters are unimportant because they are well constrained by data. Uncertainty analyses are central to the idea of a model-data feedback loop, whereby the identification of important parameters helps us prioritize additional literature synthesis and/or the collection of new data. Furthermore, PEcAn provides the required information to approach this problem quantitatively, using power analyses and economic accounting to determine the lowest cost path to reducing model uncertainty accounting for the capital (fixed) and marginal costs of research. Uncertainty analyses can also help inform data assimilation exercises both by providing highly informative priors on parameters, which restrict data assimilation to choosing biologically realistic values (that is, getting the right answer for the right reason), and by reducing the dimensionality of the data assimilation problem—parameters that have low influence can be dropped as both unimportant and unconstrainable by model-data fusion. *The combination of trait meta-analysis and data assimilation is particularly powerful as bottom-up constraints alone are not able to capture all the trade-offs and covariances among model parameters that are required to capture emergent phenomena* (Clark et al. 2014).

The PEcAn trait workflow has been applied in a growing number of applications to inform models. Dietze et al. (2014) applied PEcAn to assess the patterns in model uncertainty across 17 vegetation types and 4 biomes (figure 12.4). This analysis demonstrated that growth respiration is consistently the dominant driver of uncertainty in the Ecosystem Demography (ED) model. In subsequent analyses we split the single growth respiration parameter into a more complex submodel based on biochemical construction costs that could more directly be parameterized from trait data, thus achieving a net reduction in overall uncertainty (Shanks et al. in prep.). The next two most important parameters, stomatal slope and root conductance, are both related to plant water use—this importance of the coupling between water and carbon to overall model uncertainty has been demonstrated across a wide range of models (Schaefer et al. 2012; De Kauwe et al. 2013; Matheny et al. 2014). The fourth most important parameter, quantum efficiency, was predominantly an issue of data deficiency at high latitudes.

Beyond just using model-data feedbacks to determine important parameters, Wang et al. (2012) used PEcAn to calibrate the ED model for hybrid poplar. In this case uncertainty analysis was used to inform subsequent data assimilation, and then ensemble analysis propagated uncertainties into biofuel forecasts. Davidson (2012) used PEcAn iteratively through multiple rounds of the model-data feedback loop when calibrating ED for arctic tundra, providing an important case study in a model-informed targeted field campaign.

In addition to assimilating "pure" traits, which map directly to individual model parameters, we have also developed Hierarchical Bayesian (HB) models for assimilating data into submodels whose parameters are often treated like traits, such as the

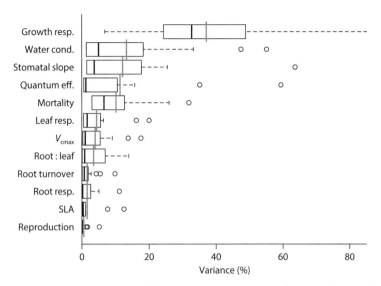

FIGURE 12.4.Uncertainty analysis of the Ecosystem Demography model. Box plot of partial variances by variable and across 17 vegetation types with the mean indicated by the vertical gray bar.

growth respiration model mentioned earlier. In Feng and Dietze (2013) we developed a Hierarchical Bayes version of the Farquhar, von Caemmerer, Berry (FvCB) (Farquhar et al. 1980) photosynthesis model and applied it to look at variations in the Vcmax-N relationship within species, across species, and across functional types over multiple seasonal cycles. This analysis demonstrated that there is not a single Vcmax-N relationship, but that the slope varies considerably depending upon scale. The scale dependence of photosynthetic parameters was also explored in Dietze (2014), which reviewed the timescales of leaf responses and explored a simple model evaluating the sensitivity of GPP to how fast leaves acclimate. This analysis illustrated (1) the importance of estimating the timescale rate constant when considering dynamic trait models (which is largely absent from the literature), and (2) the importance of circadian responses to optimizing GPP (which are largely absent from the current generation of models). In Dietze et al. (in prep) we've developed an HB meta-analysis for tree allometric relationships, which are used in ED to drive allocation, developing approaches to average over different functional relationships rather than just averaging parameters. In Shiklomanov et al. (2016), we first developed a Bayesian inversion of the PROSPECT leaf spectral model and then in Shiklomanov et al. (in prep) applied this model to look at correlations between leaf optical, economic, and photosynthetic traits, while in Viskari et al. (in review) we've explored the sensitivity of the ED model to leaf and canopy optical traits. These last three analyses form the groundwork for developing direct links between remote sensing and traits in a data-fusion framework.

To account for the multiscale correlations among trait data, as well as to combine the information provided by the various submodels with the "pure" trait data, Shiklomanov et al. (in review) has extended PEcAn's trait meta-analysis to a multivariate model. By building trait relationships into the covariance matrix, this approach does not require separate trait regressions, as the connections between all traits emerge automatically. Furthermore, these connections are bidirectional, such that not only can we use one set of traits to infer another (for example, optical traits from economic traits), but the exact same model can be used to do the reverse (for example, economic traits from optical traits) in a new context. The HB multivariate approach also automatically handles the substantial amount of missing data that occurs when extending trait correlations to larger dimensions without having to invoke additional gap-filling approaches or use different subsets of data to investigate different relationships.

Beyond model-data feedbacks PEcAn provides a wide range of other tools for processing input data, visualization, model calibration, and state data assimilation. The last of these, state data assimilation, is the basis for most real-time forecasting and will be the subject of the next two chapters.

## 12.4 KEY CONCEPTS

1. For computational and model structural reasons, less effort has gone into partitioning and propagating uncertainties in ecosystem ecology than in the other case studies.
2. I recommend that ecosystem modelers

   a. Embrace the complexities of data models and process errors in ecological systems.
   b. Fuse multiple sources of data while accounting for unreplicated sensors and systematic errors in both the data and model.
   c. Move beyond spin-up when accounting for system memory.

3. PEcAn puts the overall ecological forecasting framework (see figure 1.1) into practice for terrestrial ecosystem models, including model-data feedback loops. These feedbacks have been used to identify key parameters, reduce model dimensionality, target model structural improvements, and design field campaigns.

## 12.5 HANDS-ON ACTIVITIES

pecanproject.org > Tutorials > Hands-on Demo
Analyses on terrestrial ecosystem models:

- Trait meta-analysis
- Sensitivity analysis
- Ensemble analysis
- Uncertainty analysis

# 13

# Data Assimilation 1: Analytical Methods

SYNOPSIS: *A frequent goal in forecasting is to update analyses in light of new information. Bayesian analyses are inherently "updateable" as the posterior from one analysis becomes the prior for the next. Classic sequential data assimilation methods are designed to exploit this property to make iterative forecasts. In the first of two data assimilation chapters we focus on methods, such as the Kalman filter, that are computationally efficient and have analytical solutions, but require strong assumptions and/or significant modifications of the model code.*

## 13.1 THE FORECAST CYCLE

Where the previous chapters have described the various tools and techniques for ecological forecasting, this chapter and the next both focus on pulling these pieces together and generalizing what we've learned to make forecasts that are updatable as new data become available. Specifically, they focus on the problem of generating forecasts of some specific system state, $X_t$, at some time, $t$, in the future. Examples of forecast states might be the carbon storage of an ecosystem, prevalence of a disease, presence of a threatened or an invasive species, or stocks in a fishery.

If we just need to make this forecast once, the state-space framework (chapter 8) provides a general approach that could be used to estimate model parameters (chapter 5), partition uncertainties (chapter 6), assimilate one or more data sources (chapter 9), and project the state of the system into the future (chapter 11). Similarly, there are many ecological research questions that would benefit from hindcasting or *reanalysis*—the fusion of models and (multiple) data streams to reconstruct the past state of a system. However, a common problem is *the need to make forecasts repeatedly*, updating predictions (and/or hindcasts) as new information becomes available (chapters 3 and 4). The brute-force way to do this is to redo the whole process from scratch, performing the analysis with a combined data set of the old observations and the newly acquired data. For simple models and small data sets this approach is straightforward, but as models and data grow in complexity and volume, the obvious question is whether there's a way to update forecasts iteratively. Indeed, for many problems (both forecasting and hindcasting) it may be computationally too expensive to assimilate the entirety of the data all at once, and thus the data need to be assimilated *sequentially*, a bit at a time, in the order that they were observed.

The problem we are describing is known as the *sequential or operational data assimilation problem*, with the latter referring to the fact that this is the way that almost all real-time forecast systems operate (for example, weather forecasting). Sequential

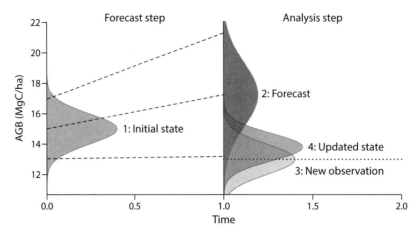

FIGURE 13.1. Forecast cycle. During the forecast step the model is used to project the initial state of the system (1)—in this case, aboveground biomass (AGB)—into the future to generate the forecast (2). Dashed lines represent mean and 95% CI. During the analysis step, Bayes' rule is used to combine the prior information provided by the forecast (2) with the information provided by new observations (3) to update the estimate of the state of the system (4). This updated state serves as the initial state for the next forecast.

data assimilation occurs through two steps: the *forecast step* projects our estimated state forward in time, with uncertainty; the *analysis step* updates our estimate of that state based on new observations. The continual iteration between these two steps is known as the *forecast cycle* (figure 13.1).

### 13.1.1 Forecast Step

The first part of the forecast cycle involves applying the error propagation approaches discussed in chapter 11 to propagate the uncertainty in our state, $X_t$, into the future. The explicit goal here is not just to propagate uncertainty through a model in general, but specifically to propagate uncertainty into the forecast (typically in time and/or space). If you recall the discussion from section 11.2, we covered a range of techniques for propagating uncertainty through a model (see table 11.1) that addressed the trade-offs between analytical (closed form math) and numerical (computer simulation) approaches as well as the choice between methods that produce full PDFs as their output versus those that returned statistical moments (for example, mean, variance). The different data assimilation methods we'll develop over the next two chapters will map directly onto these alternative uncertainty propagation choices of how the forecast step is performed. Specifically, the Kalman filter uses analytical moments for uncertainty propagation, the Extended Kalman filter uses the Taylor Series approximation, the Ensemble Kalman filter uses ensemble error propagation, and the particle filter uses Monte Carlo error propagation. The trade-offs among these methods largely parallel the trade-offs among the underlying error propagation techniques. If you have no recollection of these methods, or if you skipped chapter 11, now is the time to look back at section 11.2.

As an example of how we use these methods to make forecasts, consider the extremely simplified case of a model where the forecast ($f$) of the future state, $X_{f,t+1}$, is just a linear function of the current state, $X_t$, and we assume that the model parameters are known perfectly (no parameter uncertainty):

$$X_{f,t+1} = mX_t + \varepsilon_t$$

Simple examples of such a model would be a discrete population model, where $m$ would be the population growth rate, or a decomposition model, where $m$ would be the decay rate. For simplicity we'll also assume that the process error is Normal and additive, with mean zero and variance $q$

$$\varepsilon_t \sim N(0,q)$$

(Note: $q$ may seem like an odd choice for a variance, as will some of the other parameters here, but these are used to remain consistent with the data assimilation literature.) If there's no uncertainty about $X_t$, then our forecast is simply $X_{f,t+1} \sim N(mX_t,q)$. However, in general this isn't a great assumption so instead let's assume that the uncertainty in the state, $X_t$, is Normally distributed with mean $\mu_{a,t}$ and variance $p_{a,t}$,

$$X_t \sim N(\mu_{a,t}, p_{a,t})$$

Given such a simple linear model $X_{f,t+1}$ is also Normally distributed and we can use the analytical moments approach (section 11.2.2) to find the forecast mean, $\mu_{f,t+1}$,

$$\mu_{f,t+1} = E[X_{t+1}] = E[mX_t + \varepsilon_t] = mE[X_t] = m\mu_{a,t} \tag{13.1}$$

and variance, $p_{f,t+1}$

$$
\begin{aligned}
p_{f,t+1} &= Var[X_{f,t+1}] \\
&= Var[mX_t + \varepsilon_t] \\
&= m^2 Var[X_t] + Var[\varepsilon_t] + 2Cov[mX_t, \varepsilon_t] \tag{13.2} \\
&\approx m^2 Var[X_t] + Var[\varepsilon_t] \\
&= m^2 p_{a,t} + q
\end{aligned}
$$

which is reached by further assuming that there's no covariance between $mX_t$ and the process error. Equation 13.2 reiterates the same conclusion we reached at the end of section 11.2.3, that the uncertainty in a forecast, $p_{f,t+1}$, is dependent upon three components: the uncertainty about the initial state, $p_{a,t}$; the sensitivity of the system, $m$; and the process error, $q$. In weather forecasts $|m|$ is large (the system is chaotic), and thus reducing forecast error is achieved by minimizing $p$. By contrast, for ecological systems that are not chaotic, but rather possess some degree of internal stability, the relative importance of initial condition uncertainty tends to decline over time. Process error, on the other hand, compounds in importance over time. This conclusion highlights the importance of quantifying, partitioning, and (hopefully) explaining the process error in any ecological forecasting.

### 13.1.2 Analysis Step

Prior to observing how the future plays out, our model-based forecast, $P(X_{f,t+1})$, is our best estimate of the future state of the system. However, once the future $(t + 1)$ becomes the present $(t)$, what's our best estimate of $X_t$ conditional on making an observation of the system, $Y_t$? Naïvely, we could just use the observations themselves, $Y_t$, but doing so throws out the information from all previous observations, which may be considerable if observations are frequent or the system changes slowly or predictably. In addition, all observations are imperfect, which could easily lead to noisy forecasts if the Analysis is following the noise rather than the signal.

A better approach would be to combine our new observations with our understanding of the system and the information from our past observations, both of which are embodied in our model forecast. We can formally combine the model forecast (which is our best estimate *prior* to observing the system) with the new observations using Bayes' rule by treating the model forecast, $P(X_{f,t+1})$, as the prior and allowing the new data to enter through the likelihood, $P(Y_t|X_{f,t})$.

Continuing our simple example from before, for the data model we assume that the observed data are unbiased and have Normally distributed observation error with a known variance, $r$. Our posterior distribution is then

$$P(X_t|Y_t) \propto P(Y_t|X_{f,t})P(X_{f,t})$$

$$= N(Y_t|X_t,r)N(X_{f,t}|\mu_{f,t},p_{f,t})$$

$$= N\left(\frac{\dfrac{n_t\overline{Y}_t}{r} + \dfrac{\mu_{f,t}}{p_{f,t}}}{\dfrac{n_t}{r} + \dfrac{1}{p_{f,t}}}, \frac{1}{\dfrac{n_t}{r} + \dfrac{1}{p_{f,t}}}\right) \qquad (13.3)$$

This is the same equation as in chapter 5 (equation 5.6), when we solved for the posterior distribution of a Normal mean with known variance. As noted earlier, this solution is more easily interpreted in terms of precision (1/variance), where the mean is just a precision-weighted average between the prior (forecast) and data (likelihood). For our simplest example this is the exact solution—our posterior is the precision weighted average of the two—but that general intuition will apply more generally. So if the data are observed very precisely, the analysis will be pulled tightly to the data (figure 13.2, bottom). But if the data are observed with considerable uncertainty, we do not throw away the legacy of previous observations, but instead allow the forecast to dominate (figure 13.2, top). Furthermore, the posterior precision is the sum of the prior (forecast) and observation precisions, so the updated (posterior) estimate is always more precise than the data or model alone.

## ▓▓▓▓ 13.2 KALMAN FILTER

If we combine the forecast and analysis components of our simplest possible Gaussian linear model with known parameters, the overall forecast cycle has an analytical solution. This solution is known as the *Kalman filter*, named after Rudolf Kalman (1930–2016).

The standard version of the Kalman filter is typically generalized to the multivariate case, allowing for multiple state variables, $X$, to evolve as

$$X_{f,t+1} \sim N(MX_{a,t},Q)$$

where $Q$ is the process error covariance matrix (equivalent to $q$ earlier). The multiple state variables in the Kalman filter may represent different state variables (for example, multiple age/stage classes in a matrix population model or multiple biogeochem-

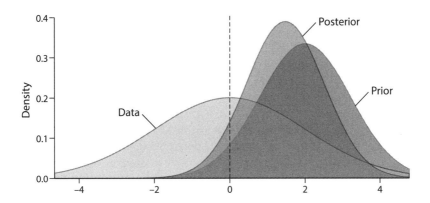

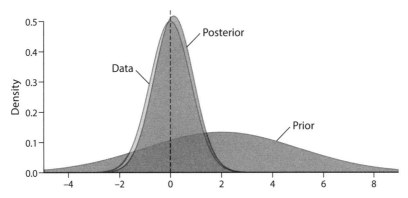

FIGURE 13.2. Impact of forecast (prior) and observation error on assimilation posterior. Top: High observation error results in an analysis posterior close to the forecast prior. Bottom: High forecast error results in an analysis posterior that tracks the observations.

ical pools), the same variable at multiple locations (for example, maps of population abundance or biomass), or multiple variables at multiple locations (for example, maps of population structure or biogeochemical pools). If we assume that the current state estimate is distributed according to a multivariate Normal with a mean $\mu_{a,t}$ and covariance $P_{a,t}$ (equivalent to $p_{a,t}$ earlier), then, *using the analytical moments approach to uncertainty propagation,* the forecast will likewise follow a multivariate Normal distribution with mean and covariance given by

$$\mu_{f,t+1} = E[X_{f,t+1}|X_{a,t}] = M\mu_{a,t} \tag{13.4}$$

$$P_{f,t+1} = Var[X_{f,t+1}|X_{a,t}] = Q_t + M_t P_{a,t} M_t^T \tag{13.5}$$

This solution is directly equivalent to the univariate solution (equations 13.1 and 13.2) with the only notable difference being that $m^2 p_{a,t}$ translates to the matrix equivalent $M_t P_{a,t} M_t^T$ (not $M_t^2 P_{a,t}$), where the superscript $T$ indicates matrix transpose.

The data model in the Kalman filter assumes that we can make multiple observations, $Y$

$$Y_t \sim N(HX_t, R)$$

where $R$ is the observation error covariance matrix, and $H$ is known as the *observation matrix*, which in the simplest form simply *maps a vector of observation to which state variables they match*. If we have $v$ observations and $n$ states, then $Y$ is a $[v \times 1]$ vector, $X$ is a $[n \times 1]$ vector, and $H$ is a $[v \times n]$ matrix of 1's and 0's specifying which observations (rows) go with which variables (columns). Similar to section 9.2, there is no requirement for which is bigger, $v$ or $n$, nor is there a requirement that all states need to be observed. More generally, the observation matrix, $H$, does not strictly have to contain just 0's and 1's but can also be used to transform the model states to be equivalent to the observations, which we have seen previously in other data models, such as the calibration curve example in chapter 6. $H$ can also be used to address spatial misalignment (section 9.3). For example, one remotely sensed pixel ($y_i$) may represent the weighted mean of what's observed across multiple model grid cells, in which case that row, $H_{i,\cdot}$, would contain the weights.

As we did with the univariate model, we can again use Bayes' rule in the analysis step to update our estimate for the state vector $X$

$$X_{a,t}|Y_t \propto N(Y_t|HX_{a,t},R)N(X_{a,t}|\mu_f,P_f)$$

$$= N\big((H^TR^{-1}H + P_f^{-1})^{-1}(H^TR^{-1}Y_t + P_f^{-1}\mu_{f,t}),(H^TR^{-1}H + P_f^{-1})^{-1}\big) \qquad (13.6)$$

This solution looks complicated, but it is simply the matrix equivalent to equation 13.3. If you look closely, you'll see that the term $(H^TR^{-1}H + P_f^{-1})$, which shows up both on the left-hand side of the mean and in the variance, is simply the sum of the observation, $R$, and forecast, $P_f$, precisions. On the right-hand side of the mean we see that $(H^TR^{-1}Y_t + P_f^{-1}\mu_{f,t})$ is just the weighted average of the data, $Y$, and the forecast (prior), $\mu_f$. To make all this more clear equation 13.6 is often simplified to

$$\mu_{a,t} = E[X_{a,t}|Y_t] = \mu_{f,t} + K_t(Y_t - H\mu_{f,t})$$
$$P_{a,t} = Var[X_{a,t}|Y_t] = (I - K_tH)P_{f,t}$$

where $K$ is known as the Kalman Gain and is equal to

$$K = P_{f,t}H^T(R + HP_{f,t}H^T)^{-1}$$

As with the univariate case, the Kalman Gain makes it clear that the updated, posterior estimate is the weighed mean of the model and the data. The Kalman Gain formulation is also more computationally efficient (see box 13.1) because it involves fewer matrix inversions and, when $v < n$ (number of observations per time point < number of states), can greatly reduce the size of the matrix being inverted.

One of the powerful features of the multivariate Kalman filter is that it allows us to make inferences about unobserved state variables based on the covariance between the observed and unobserved variables. For example, imagine that a model predicts a strong covariance between the aboveground biomass on a landscape and the fraction of that landscape that is covered with conifers (figure 13.3, left). We might be able to measure, with uncertainty, the fraction of the landscape that is conifer, for example, using remote sensing, field observations, historical records, or fossil pollen. In the analysis step we can use Bayes' rule to update our estimate of the conifer fraction (figure 13.3, center). Furthermore, by exploiting the covariance between conifer fraction and biomass, we also update our biomass estimate (figure 13.3, right), with the amount of constraint determined by the strength of the correlation. Interestingly,

## Box 13.1. Localization

Large covariance matrices are not uncommon throughout statistics. In this book we've seen them show up first when dealing with autocorrelation in time and space (chapter 6), where the size of the matrix ($n \times n$) was determined by the number of observations ($n$). We've now encountered it again in the analysis step of data assimilation, where $n$ is the number of state variables, $X$, in the system. If we are trying to model multiple state variables across spatial domains with many grid cells, then $n$ can easily become very large. As noted in the text, matrix inversion is an $O(n^3)$ problem, so for example if you needed a higher resolution map and you wanted to halve the size of the grid cell used (for example, going from $30 \times 30$ m to $15 \times 15$ m), this increases the number of grid cells by 4×, which increases the computation by 64×. Fortunately, there are alternative approaches to matrix inversion that can be considerably faster if the matrix is sparse (lots of 0's) or if it is highly structured (for example, occurring in blocks or diagonal bands) (Press et al. 2007). Furthermore, code libraries exist in all major programming languages dedicated to these types of matrix inversion and linear algebra problems. The practical problem then becomes how to structure your problem to take advantage of these algorithms. This is typically achieved by *localization*, which involves *assuming certain correlations are zero*. For spatial problems this typically involves setting correlations to zero beyond some threshold distance, which is usually determined by initial exploratory analysis looking at how spatial autocorrelation decays with distance. Note that while the distance involved in localization thresholds is typically spatial, there is no requirement on what distance metric is used in localization and any ecological distance would be allowed (for example, phylogenetic, community similarity, bioclimatic, spectral, and so on). Also remember that this distance metric is not used in the calculation of the covariance between two points, which is the covariance determined by the model and is not an explicit function of any distance measure. Rather, this metric is simply used to determine whether that covariance is calculated at all or assumed to be zero.

Choosing a threshold distance involves a trade-off between computation and information. Making the threshold distance as small as possible sets more of the matrix to zero and thus decreases the computational cost. Making the threshold too small limits the spatial range over which observations are allowed to inform the model. Localization can also be used to limit the cross-correlations among state variables, limiting how one state variable impacts other state variables, both within a location and as a function of distance. Indeed, in some cases correlations among variables may be judged to be spurious, in which case localization not only increases computational efficiency but also improves the science. Finally, dealing with covariance matrices efficiently requires not only localization, but also that you structure $X$ intelligently to keep variables organized in ways that matrix algorithms can work with efficiently, such as large bands and blocks. This is typically achieved by organizing $X$ first by response variable (which introduces block structure to the covariance) and

*(Box 13.1 cont.)*

then by distance to keep nearby locations together (which keeps covariances close to the diagonal band). For example, if the case study in figure 13.3 were to be extended across space, one might sort $X$ such that all the fraction conifer forecasts came first, and then all the biomass forecasts came second. Then within each block, locations could be sorted by distance. Finally, a distance threshold would be applied to set location pairs that are "far away" to zero.

the math for estimating the unobserved variable(s) is identical to *Kriging* (a classic geostatistical form of spatial interpolation based on spatial autocorrelation [section 6.6]), but here the covariance matrix is provided by the model structure rather than the spatial autocorrelation. An important caveat to this approach to data fusion, which is the same caveat for any form of data fusion (chapter 11), is that we're assuming that the covariances in the model are structurally correct, and that the model has been calibrated and validated. If a model is structurally incorrect, then assimilating one class of data will lead to structural biases in our estimate of the other. That said, if the model is structurally incorrect, it's already generating biased forecasts, and data assimilation will not solve that problem for latent variables. In other words, for an observed quantity data assimilation leads to a tighter constraint on both the accuracy and the precision, while for an indirectly inferred quantity (that is, a latent variable), it leads to more precise estimates (lower uncertainty) without any guarantee that such estimates are more accurate.

Data assimilation is ultimately a form of data-driven inference where model structure (the embodiment of process understanding) provides the covariance matrix. Furthermore, even when observations are not absent, the Kalman filter can also be a tool for data synthesis (chapter 9), leveraging model structure to be able to combine observations of different variables that may operate at different spatial and temporal

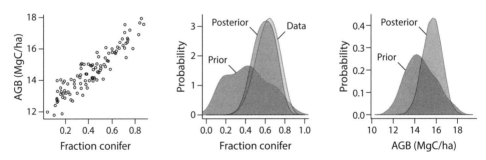

FIGURE 13.3. Joint constraint of observed and unobserved quantities. Envision an ensemble of model forecasts (left) that exhibit a covariance in the predictions for aboveground biomass (AGB) and fraction of the landscape that is covered with conifers. The marginal distributions of this joint probability are summarized as the prior distributions in the center and right panels. If the fraction conifer is observed with uncertainty (center panel, Data) then Bayes' rule can be used to update the model estimate (center panel, Posterior). Furthermore, the unobserved AGB is also updated because of the covariance between the two states (right panel, Posterior). Figure reproduced from Dietze et al. (2013).

scales. However, this idea of "models as a scaffold" for data synthesis (see figure 1.3) is not unique to the Kalman filter but is a property of the more general state-space framework (chapter 8), of which the Kalman filter is just one special case (Dietze et al. 2013).

### 13.3 EXTENDED KALMAN FILTER

As noted earlier, the Kalman filter (KF) is a special case of the state-space model where the model is linear and both the process and observation errors are Normally distributed. It did not take long for the limitations of these assumptions to become apparent. Historically, the first assumption that was relaxed was that of linearity in the process model. The Extended Kalman filter (EKF) was developed to allow the process model to be an arbitrary nonlinear function, $X_{t+1} = f(X_t)$. In this case the Kalman analysis step, with its assumption of Normal data and a Normal forecast, stays the same as the original Kalman filter, but the update of the forecast mean is done using the full model. Furthermore, in general the forecast variance cannot be updated with the fully nonlinear model, so instead it is *approximated with a linear model using the Taylor Series approach* (section 11.2.3). In the univariate version, this means setting $m$ in equation 13.2 to the first derivative of the process model, $m = df/dX$. In the multivariate version this leads to a solution involving the Jacobian, $F$ (*the matrix of all pairs of first derivatives, $\partial f_i/\partial X_j$*), in place of the matrix of slopes ($M$) in equation 13.5.

Importantly, in the EKF the Jacobian is evaluated around the current mean state *at each timepoint*, so the values in the matrix $F_t$ do not stay constant through time. This adds one more step to every iteration of the forecast step: first, the current mean of $X$ needs to be plugged into the equations that result from solving for the derivative(s) to get specific numerical values for $m$ or $M$; second, you update the forecast mean and variance. As with any numerical integration, if the forecast is for more than just the immediate future, then in practice you will want to iterate between these two time steps over some finite time step (Press et al. 2007).

To give a concrete example of the EKF, consider the logistic growth model

$$N_{t+1} = N_t + rN_t\left(1 - \frac{N_t}{K}\right) \tag{13.7}$$

which has a first derivative that depends on $N_t$

$$\frac{\partial N_{t+1}}{\partial N_t} = 1 + r - \frac{2r}{K}N_t \tag{13.8}$$

If we assume, for example, that $r = 0.3$, $K = 10$, and $N_0 = 3$ with a standard error of $p_{a,0} = 1$, then our forecast for the first time step is

$$m_1 = 1 + 0.3 - \frac{2 \cdot 0.3}{10}\,3 = 1.12$$

$$N_1 = 3 + 0.3 \cdot 3\left(1 - \frac{3}{10}\right) = 3.63$$

$$p_{f,1} = m_1^2 p_{a,0} + q \approx 1.25 + q$$

We would have to specify a process error, $q$, to further evaluate the function. If we want to make a forecast for the second time step, but we haven't yet made any new

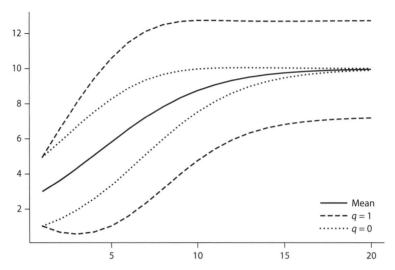

FIGURE 13.4. Extended Kalman filter forecast (mean and 95% CI) of logistic growth under alternative choices of process variance, $q$. The $x$-axis is time, and the $y$-axis is the state variable, $X_t$.

observations at the first, we can just skip the analysis (since the posterior equals the prior if there's no likelihood) and iterate the preceding steps again, updating $m_2$, $N_2$, and $p_{f,2}$ from $m_1$, $N_1$, and $p_{f,1}$. Figure 13.4 illustrates an example of running this EKF forecast to steady state under two alternative values of process variance, $q = 0$ and $q = 1$. In the first scenario ($q = 0$) uncertainty declines asymptotically to zero, while in the second ($q = 1$) the total uncertainty increases until $t = 5$ and then declines asymptotically to reach a steady state. The $q = 0$ scenario can also be interpreted as the contribution of initial condition uncertainty to the $q = 1$ scenario. Initial condition uncertainty dominates initially but is overtaken by process variance. The difference between the interval estimates between $q = 0$ and $q = 1$ thus also illustrates the contribution of process error to the second scenario.

The KF and EKF are extremely widely used approaches beyond just weather forecasting. Indeed, one of the first applications of the KF was as part of the navigation system in the Apollo moon landing. KF and EKF models are likewise embedded in most GPS systems and are commonly used for digital object tracking.

The KF and EKF have a number of advantages over the more general state-space framework, most important being that having an analytically tractable solution makes them computationally efficient as well as providing a more general understanding. Furthermore, the forecast/analysis cycle provides an updatable approach where the forecast is dependent only upon the current state and the analysis depends only upon the forecast and the current observations. This is a stark contrast with full MCMC approaches that depend upon the full history of all observations to update the forecast when new observations are made. All these advantages inevitably come at a cost. For example, the fact that inference depends only on the most recent state means that current observations do not provide any retroactive constraint on recently estimated states, which is in contrast to the state-space models we saw earlier (chapter 8). The KF and EKF also achieve their efficiency through strong assumptions of Normality and either a linear model or a local linearization. For the EKF there is also the cost of needing to analytically or numerically solve for the full Jacobian matrix, which can be daunting for large models. In addition, for models that have many state variables, such as those making spatial projections across a grid, the computational cost of inverting the covariance matrix can be high (see box 13.1). Computer scientists classify standard matrix inversion as an $O(n^3)$ problem (read as "Order $n^3$"), meaning that the computation cost increases as the cube of the size of the matrix. This means that the computational costs go up very quickly as the number of state variables or locations increases.

Finally, the KF and EKF make the assumption that all the parameters are known, which is usually a much less tenable assumption in ecological forecasting problems than more physical applications. It is worth noting that *it is possible to augment the vector of state variables, X, by including model parameters as additional unknowns.* Indeed, this approach has been applied in a number of ecological models to estimate model parameters (Fox et al. 2009). This approach can be particularly handy for large models where brute-force MCMC techniques can be computationally prohibitive. However, while it is possible to augment $X$ with parameters that are part of the deterministic part of the model, it is not possible to use this approach to estimate the observation error or process error. While it is sometimes possible to generate independent estimates of observation error, in general it is much harder to estimate the process error in ecological problems, and this term often ends up being hand tuned. This is an unsatisfying approach given that the results of data assimilation are quite explicitly the weighted average between the uncertainty in the data versus the model and we expect the process error to dominate in the long term. In the following chapter we will look at alternative numerical approaches to further relax the assumptions of KF and EKF.

## 13.4 KEY CONCEPTS

1. A common problem is the need to make forecasts repeatedly, updating predictions as new information becomes available.
2. Sequential data assimilation occurs through two steps: the forecast step projects our estimated state forward in time, with uncertainty; the analysis step updates our estimate of that state based on new observations. The continual iteration between these two steps is known as the forecast cycle.
3. The forecast step involves applying the error propagation approaches discussed in chapter 11 to propagate the uncertainty in our state, $X_t$, into the future. The different forms of data assimilation discussed in this book map directly onto the alternative choices of how the forecast step is performed.
4. The analysis step uses Bayes' rule to combine the model forecast (which is our best estimate *prior* to observing the system) with new observations (through the likelihood).
5. The Kalman filter (KF), which combines a linear model with known parameters and Gaussian process and observation errors with known uncertainties, is the simplest example of the forecast cycle. Using the analytical moments approach it has an *iterative* analytical solution.

6. Data assimilation allows us to use our model as a scaffold for data-driven synthesis, making inference about unobserved state variables based on the covariance between the observed and unobserved variables.
7. The Extended Kalman filter (EKF) relaxes the assumption of a linear model by applying the Taylor Series approach to uncertainty propagation in the Forecast step.

## 13.5 HANDS-ON ACTIVITIES

https://github.com/EcoForecast/EF_Activities/blob/master/Exercise_09_KalmanFilter .Rmd

- Kalman filter

# 14

## Data Assimilation 2: Monte Carlo Methods

*Synopsis: In the previous chapter we introduced the concept of the forecast cycle and derived the classic Kalman filter (KF) and Extended Kalman filter (EKF) based on analytical approaches to uncertainty propagation (linear transformation and Taylor Series respectively). In this chapter we extend these approaches using the numerical methods for error propagation discussed in chapter 11, ensemble and Monte Carlo methods, to the data assimilation problem. This leads to the Ensemble Kalman filter and the particle filter, respectively. We then discuss the generalization of these approaches to the ecological forecasting problem. Finally, we touch on the issue of forecasting with multiple models, discussing approaches such as Bayesian model averaging and reversible-jump MCMC.*

### 14.1 ENSEMBLE FILTERS

As noted in the previous chapter, the Extended Kalman filter (EKF) was derived as a means to linearly approximate the forecast step of the forecast cycle when working with nonlinear models. This approach is essentially replacing the analytical moments approach to uncertainty propagation discussed in chapter 11 with the Taylor Series linear approximation approach. In discussing uncertainty propagation we also presented two numerical methods: ensembles and Monte Carlo. In the *ensemble-based approach* the size of the sample is typically large enough to allow the accurate estimation of the mean and covariance, but not so large as to represent the full probability distribution. This same approach can be applied in the forecast cycle. Specifically, the joint posterior distribution from the analysis step is sampled, and *the forecast step is performed by running the model forward independently for each ensemble member* (figure 14.1). The result is an ensemble-based filter. The basic steps involved in an ensemble filter are thus:

FORECAST STEP

1. Jointly sample state variables from their current joint distribution (previous analysis posterior or, at the start, from the initial condition prior), where $m$ is the size of the ensemble and $X_{a,i}$ is the $i^{th}$ member of the ensemble.
2. For each ensemble member, independently run the model into the future to generate the $i^{th}$ forecast, $X_{f,i}$.
3. Fit a *known* probability distribution to the ensemble output, $X_{f,*}$. This distribution serves as the forecast (prior) distribution for the analysis step and

is typically done using the sample mean and covariance from the ensemble. In the classic Ensemble Kalman filter this would be a multivariate Normal.

ANALYSIS STEP

1. Use analytical or numerical techniques to combine the forecast (prior) with the data model (likelihood) to generate the analysis (posterior). This step is no different from any other Bayesian statistical model.

Compared to the EKF, this approach has the advantage of being relatively easy to implement, as it does not require an analytical Jacobian (matrix of first derivatives) or any changes to the underlying model structure or code. It is also easy to understand, and capable of dealing with a high degree of nonlinearity. Since it is using an ensemble-based method, we do not need to worry about the impact of Jensen's inequality on the forecast statistics (for example, mean, variance, CI), which was an issue for the analytical methods (for example, EKF). The ensemble-based approach also makes it easier to accommodate other sources of uncertainty, such as uncertainty in parameters and model drivers, or stochastic processes such as disturbance. These can be included by sampling them as well, at the same time as the sampling of the initial conditions, thus producing an ensemble forecast that includes multiple sources of uncertainty simultaneously. The disadvantage of the ensemble approach is that it requires the additional computation of a larger number of model runs, rather than running the model once (as was done in KF and EKF). All in all, these trade-offs have made ensemble forecasting one of the most popular data assimilation approaches in ecology.

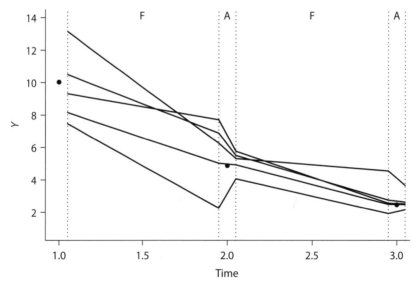

FIGURE 14.1. Ensemble Kalman filter (EnKF) applied to a simple linear model ($m = 0.5$, $q = 1, r = 1$). At the start the ensemble is sampled from the initial condition PDF. During the forecast step (F), each ensemble member is independently run forward using the model, with stochasticity introduced via the process variance. During the analysis step (A) the ensemble is adjusted based on information provided by the observations (solid black dots).

The most common ensemble-based approach for approximating the forecast step is the *Ensemble Kalman filter (EnKF)* (Evensen 2009a, 2009b). EnKF retains the Normality assumptions of the Kalman filter, and thus the analysis step is identical to the Kalman filter (equation 13.6). The forecast (prior) distribution is likewise approximated as a Normal—in this case, using the *ensemble forecast sample mean and sample covariance* instead of an analytical projection of the forecast mean and covariance, as was done in the KF and EKF

$$\mu_{f,t+1} \approx \frac{1}{m}\sum X_i$$

$$P_{f,t+1} \approx COV[X]$$

While the EnKF does not violate Jensen's inequality when transforming nonlinear models, it does unfortunately violate its own Normality assumption, just as the Extended Kalman filter (EKF) did. Specifically, if we take a sample from a Normal analysis posterior and perform a nonlinear transformation, we know that what comes out the other end can't actually still be a Normal distribution. As with the EKF, the degree of nonlinearity will determine just how bad an assumption the Normal forecast prior turns out to be. In the ensemble-based filter, the size of the ensemble scales with the dimensionality of the model (that is, more state variables means a larger ensemble is required to estimate the covariance matrix) and the degree of nonlinearity in the model (highly nonlinear models need larger ensembles). In particular, unless localization is employed (see box 13.1), the number of ensemble members has to be greater than the number of state variables to ensure the covariance matrix is invertible. In practice the exact sample size should be chosen based on a power analysis (section 11.4.1) to determine how the numerical accuracy (for example, standard error) in the approximation scales with the ensemble size.

While the EnKF is the most common form of ensemble-based data assimilation, the analysis step is conceptually easy to generalize to any arbitrary choice of forecast prior and likelihood in the analysis step. The only requirement is that the forecast prior must be a named distribution that is fit based on the sample of ensemble runs. If the forecast is multivariate, then this named distribution must likewise capture the full joint prior distribution, not just the marginal distributions, since, as discussed in the previous chapter, the covariances among parameters are an essential part of data assimilation. The analysis step can involve any non-Gaussian data model as well. That said, when generalized to an arbitrary likelihood and forecast prior, the analysis is unlikely to have an analytical solution (unless the distributions were deliberately chosen to be conjugate). However, as we learned in chapter 5, MCMC methods can be used to sample from most any analysis posterior and, given the nature of the ensemble-based method, that sample is exactly what we need to initialize the forecast step anyway. In practice, the popularity of the EnKF formulation specifically is likely because (1) the Normal often is a good approximation, and (2) the specification of the multivariate covariance structure in the forecast prior can be challenging in the general non-Normal case. It is also likely, due to the scarcity of non-Gaussian ensemble filters in the literature, that some users do not realize that the EnKF can be easily generalized.

As with the more generic problem of ensemble error propagation, the computational burden of ensemble forecasting can sometimes be reduced by choosing the ensemble members systematically rather than randomly. For example, if we propagate uncertainty in the forecast using the Unscented Transform discussed in chapter 11, this leads to a form of the EnKF known as the Unscented Kalman filter (Uhlmann 1995; Julier et al. 2000; Gove and Hollinger 2006) (box 14.1).

## 14.2 PARTICLE FILTER

In the EnKF, we forecast with an ensemble that is large enough to estimate the mean and covariance, but not large enough to approximate the distribution. This raises a question: *could we instead do a large Monte Carlo forecast?* As with Monte Carlo error propagation (chapter 11) and the MCMC approach (chapter 5), by using a large Monte Carlo sample we can eliminate the need for any closed-form analytical solution for the posterior, and instead approximate the distribution with the posterior samples. This solves the problem, in the EnKF and EKF, that the Normal forecast prior has to be a false assumption, and can even allow for forecasts with multiple modes (for example, one mode for ensemble members that continue along their current trajectory and another for those where disturbance occurred, or which shift to some alternative steady state). However, the question then becomes, how do we perform the analysis step if our prior is just a sample?

The particle filter (PF), also known as *sequential Monte Carlo (SMC)*, provides a means for updating a Monte Carlo *prior* to generate updated Monte Carlo *posterior* samples (Doucet et al. 2000; Del Moral et al. 2006; Doucet and Johansen 2011). The fundamental idea is to *weight the different ensemble members (referred to here as particles) based on their likelihood during the analysis step.* Initially, when we take a random sample from a distribution (as we do in any MC analysis) each particle has the same weight ($w = 1$). Therefore, by multiplying that weight of 1 by the likelihood, the updated weight ($w = L$) is performing our standard Bayesian calculation of likelihood times prior. Furthermore, the sum of these weights provides the normalizing constant for the denominator in Bayes' rule (chapter 5). Given this set of weights, we can then calculate all of our standard metrics (mean, variance, quantiles, histogram, empirical cumulative distribution function [CDF], and so on) based on the weighted version of these calculations. For example, the weighted mean and variance of $X$ are just

$$\mu = \frac{\sum w_i X_i}{\sum w_i}$$

$$\sigma^2 = \frac{\sum w_i (X_i - \mu)^2}{\sum w_i}$$

The PF approach can also be applied iteratively, with the posterior of one analysis forming the prior for the next, by simply continuing to multiply the likelihoods from each stage of the analysis ($w = L_1 L_2 L_3 \ldots$). With each subsequent round of analysis we are downweighting unlikely variable combinations and upweighting the likely ones. While this is exactly what we want, this approach runs into a few practical problems, the dominant one being that the bulk of the weight can quickly converge onto the small subset of the particles that has the highest likelihood. As this begins to happen, the effective number of particles in the MC approximation can decrease rapidly, leading to very coarse estimators. Indeed, it is not unheard of for almost all the weight to end up on a single ensemble member.

## Box 14.1. Filter Divergence and Ensemble Inflation

Practitioners of data assimilation in the atmospheric sciences frequently worry about a problem known as *filter divergence*. Filter divergence occurs when the combination of inherent variability and processes error in a model are small, which leads to low forecast uncertainty. On face value this sounds like a good thing, but if the forecast uncertainty is considerably less than the observation error ($p_f \ll r$), the model always "wins" during the Bayesian updating, and over time the filter then ignores (diverges from) the data (figure 14.2). In the extreme case, with a stable model and no process error ($q = 0$), the filter will converge asymptotically to a point estimate, even if the data exhibit a change to a very different system state (for example, disturbance, alternative steady state, and so on).

In atmospheric science the solution to filter divergence has traditionally been to hand-tune the process variance to a larger value to prevent divergence from happening, a process sometimes called *ensemble inflation*. Tuning is obviously unsatisfying statistically, since the process variance has a large impact on the forecast, so it really should be estimated from data using the techniques discussed in chapters 6 and 8 and section 14.4. Tuning is also unsatisfying ecologically since high process variability often means that important unknown factors are driving system dynamics. Understanding and partitioning such variability can be key to advancing our science.

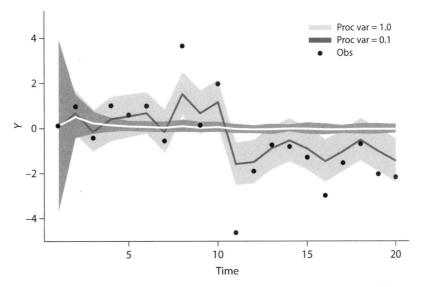

FIGURE 14.2. Filter divergence. Two alternative fits of an EnKF to simulated data (black dots) with known parameters ($m = 0.5, q = 1, r = 1$) with either the correct process error ($q = 1$, light gray CI and line) or too low process error ($q = 0.1$, dark gray interval and white line). In the low scenario the model converges tightly on the model prediction and ignores (diverges from) the observed data.

*(Box 14.1 cont.)*

Compared to atmospheric models, ecological models are even more likely to exhibit filter divergence because they tend to have more stabilizing feedbacks and less chaos. This raises an important, and unanswered, question as to whether filter divergence is always the wrong answer ecologically. Many ecological systems have long memory and therefore we should expect to be more confident in our understanding of a specific site or process as we continue to measure it (provided that variability, autocorrelation, bias, and so on are handled correctly; see chapter 9). If the model and data then diverge, this could indicate that something is wrong with the process model or the partitioning of process error. In this case, inflating the variance would just mask an error. Instead, stopping to fix things (chapter 4) when filter divergence occurs is the correct thing to do (though this requires running diagnostics to identify such an error so you know to stop; see chapter 16).

By contrast, there will always be things that the models do not account for, as ecosystems are subject to a wide range of low probability events that don't always make sense to try to model explicitly. In this case we typically want the data assimilation system to respond to these system changes. However, that can't occur unless the process model explicitly includes some low but nontrivial probability of a system-state change. This doesn't require inflating the variance as much as it requires inflating the tails of the prediction or introducing a prediction with multiple modes (for example, intact versus disturbed). Overall, other than avoiding hand-tuning, the issue of filter divergence in ecological forecasting is not black and white. The important lesson is to keep an eye out for this issue and to take the time to diagnose what is going on when it is observed.

The standard solution to this problem is to *resample* from the particle filter based on the weights (figure 14.3). Individual ensemble members (also known as particles) that have high weight will be resampled more frequently than members with low weight, many of which will be dropped. In essence, the ensemble members that are resampled more than once are cloned multiple times into new ensemble members. For this approach to work, however, requires that the ensemble members not be deterministic—there needs to be some underlying source of variability (for example, process error) that causes the cloned ensemble members to diverge from one another; otherwise, the filter will still converge to a point estimate. In most cases this means adding process error onto each ensemble member in the forecast step. Furthermore, it is important to note that *once the ensemble members are resampled the weight assigned to each particle is reset to 1*. The information that was previously captured by the weights is now captured by how many times an individual particle was cloned.

Part of the art of the particle filter is determining when to resample. If done too often (for example, every time new data are observed), we may lose particles simply through drift, and once a particle is lost from the ensemble, it can't be entered back in. As a simple example, imagine the particles are the numbers 1 through 10. If we resample these at random (equal weight), some of them will inevitably be lost, say 5

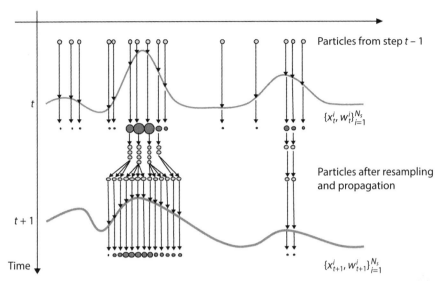

FIGURE 14.3. Particle filter. At the start of the analysis (which represents either the initial prior or the ensemble output from the forecast step) all particles all have equal weight, which is graphically indicated by their equal size (top row of circles). At time $t$ the likelihood (curve) is used to assign a weight to each particle based on the probability of the data given that particle's state, which is indicated by the size of the particle below the curve. Particles may then be resampled in proportion to their weights (vertical columns of equal-sized particles). This completes the analysis step. During the forecast step each particle is then run through the model independently, causing individual particles to diverge from one another, which is indicated in the graph by the nonvertical arrows below the vertical columns of particles. The process is then repeated at the next time step $(t + 1)$. Reproduced from Lee et al. (2012).

and 8, while others are duplicated, say 3 and 7. If we then resample from the resulting set, we might lose more numbers and duplicate others. If this is done repeatedly, then eventually *we will converge on a single number through drift*, even if all particles were weighted equally. At the other extreme, if we do not resample often enough, then we return to our original problem of the weight converging on too small a subset of particles. This problem is typically addressed through simple heuristics that decide when to resample—for example, when the largest weight exceeds some threshold or when the cumulative weight is focused on too small a subset of the particles.

There are many advantages to the particle filter. As with ensemble filters, the PF is relatively easy to implement; however, it is even more general and more flexible as it does not enforce distributional assumptions on the prior or posterior. Therefore, it is also able to capture complex covariance structures without requiring Gaussian data models. Like with the Kalman filter variants, in the particle filter it is possible to augment the variables being estimated with the model parameters and error parameters in addition to state variables. However, care must be taken when using the particle filter to estimate parameters and variances as these components *are generally treated as deterministic*—they do not change in value over time or have process error the way the model states do. Therefore both drift and convergence become bigger issues for these terms, though additional sampling techniques do exist to try to balance this trend (Doucet et al. 2000; Del Moral et al. 2006; Doucet and Johansen 2011). The obvious disadvantage of the particle filter, as a whole, is the considerably larger computation required. That said, both ensemble and particle filters have the advantages of being simple to parallelize, as different ensemble members can be run on different CPUs.

## 14.3 MODEL AVERAGING AND REVERSIBLE JUMP MCMC

In all of the preceding discussions of ensemble-based forecasting we have assumed that these forecasts are being made with a single model and we're trying to account for the uncertainties within that model. However, this ignores a critical source of uncertainty, which is *model structure*. Very rarely in ecology do we believe that the equations we've written down are the "correct" model. Indeed, it is very often the case that different alternative model structures exist within a community, either as entirely different models or as alternative variants within a single model.

The most common way that model structural uncertainty is accommodated is by making forecasts using an ensemble of different models (chapter 12), each of which makes different assumptions. Given a discrete set of models, one can then average over their forecasts, and indeed this multi-model average very frequently performs better than the individual models that make up the ensemble forecast (Weigel et al. 2007; Schwalm et al. 2010). In the simplest case, all models are counted equally in an ensemble forecast. Better yet is when each of the models also propagate their own uncertainties—for example, by generating an ensemble of ensembles. That said, at times we may not want to count all models equally, but instead give weights to different models based on their past performance. In this case *Bayesian model averaging (BMA)* provides a formal way to combine multiple models when making forecasts (Hoeting et al. 1999; Raftery et al. 2005). The forecast in BMA then becomes a weighted average of the individual forecasts from each model. Since those forecasts are typically posterior distributions, not point values, the combined forecast is then a mixture distribution (see figure 6.2). In some variants of model averaging the weights given to models do not have to be static but can evolve through time as new data are encountered (that is, data assimilation can update the weights) to give more weight to models that have been predicting well.

For the alternative case, where structural uncertainties are accounted for by variation within a single model, reversible jump MCMC allows for multiple nested models to be fit simultaneously (Hastie and Green 2012). Within this approach multiple models are included within a single MCMC. For example, the sampler would update not just the $k^{th}$ parameter value, $p(\theta|k)$, but also whether the $k^{th}$ term should be included in the model, $p(k)$. From iteration to iteration the number of parameters changes, with parameters blinking in and out from the model as it mixes. In the end reversible jump MCMC returns a posterior probability for each alternative model, in addition to the parameter estimates, and is useful for evaluating alternative structures within a single model. The probabilities of alternative models can then be used either as a form of model selection, eliminating models with very low probability, or as weights in model averaging.

Finally, when combining multiple models it is important to note that the set of models in hand rarely represents a random subset of all possible models for a process.

In addition, in many communities the models themselves are not independent, having often evolved from the same parent model or having adopted structures from other models in the community. Indeed, while this lack of independence is often much bemoaned for the statistical problem it introduces, in reality it is not always a bad thing if it means that different teams are learning from the successes and failures of their colleagues. Still, this lack of independence and lack of random sampling must always be acknowledged when working with ensembles of multiple models (Tebaldi and Knutti 2007).

## 14.4 GENERALIZING THE FORECAST CYCLE

If we return to the ideas we presented at the start of chapter 13, when we introduced the forecast cycle, the key insights are (1) that the analysis step is just an application of Bayes' rule (chapter 5) with the forecast as the prior, and (2) that the forecast step is just uncertainty propagation (chapter 11). If we look at the different data assimilation approaches presented in the last two chapters, they are mostly illustrating how we can apply the different uncertainty propagation approaches (KF = analytical moments; EKF = Taylor Series; EnKF = ensemble; PF = Monte Carlo) to the forecast step. The pros and cons of different methods largely parallel (and fall out of) the trade-offs summarized in table 11.1: numerical methods (for example, EnKF, PF) are simpler, more flexible, and, for nonlinear models, generally more accurate, but are computationally more expensive than analytical approaches (KF, EKF). The computational cost typically increases the more a model violates the assumptions of Normality and linearity, but this is precisely where these methods are most needed, thus setting up a further tension between computation and accuracy. For example, Monte Carlo error propagation (PF) involves the fewest assumptions and provides the most detailed output (samples from the full joint PDF), but is generally the most computationally expensive.

More generally, all of these data assimilation approaches are at their heart just special cases of the state-space model (chapter 8), where you have some dynamic process model, $f(x)$, evolving the latent process, $X$, through time and some observation model, $g(x)$, which links that latent state to observations. As such, the overarching framework of the forecast cycle can be generalized to deal with the complexities of real data (chapter 6), to combine multiple data sources (chapter 9), and to use any uncertainty propagation approach (chapter 11).

In generalizing the forecast cycle, the full state-space model is the "gold standard" but is computationally demanding because we typically solve it using MCMC. The approaches described here (KF, EKF, EnKF, PF) are based on specific special cases to the state-space model, and the stronger the assumptions we are willing to make, the more we can speed up the computation. In general, the Kalman filter made the strongest assumptions (linear, Normal) and each subsequent methods relaxed these assumptions. However, it is important to note that *all* of the approaches discussed in these two chapters involve simplifications aimed specifically at dealing with the Initial Condition problem for chaotic models with known parameters. In most cases, this is not the Ecological Forecasting problem. So what simplifying assumptions to the general state-space framework are appropriate for ecological forecasting? This remains an important open question. As a whole, the goal of this book has been to provide the reader with the general framework and individual tools and techniques

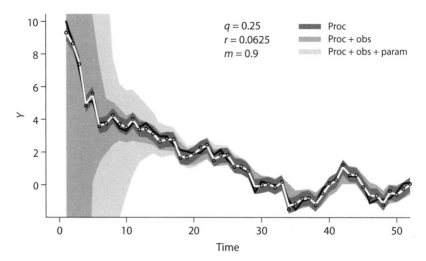

FIGURE 14.4. Generalization of the univariate Kalman filter to unknown process variance (Proc, $q$), unknown observation variance (Obs, $r$), and unknown model parameters (Param, $m$). Shaded regions represent 95% CI for three different scenarios sequentially adding different unknown terms. This approach is applied to a simulated time series (black line) with known parameters from which pseudo data were generated with observation error (open circles). White line indicates median fit under process variance scenario.

to be able to answer that question for the specific forecasting problems faced by different communities.

One particular problem that is pervasive in ecological forecasting is that there are many sources of uncertainty beyond just initial conditions, and it is quite likely that for many problems other sources, such as process error, will dominate. However, none of the Kalman filter variants presented earlier can be used to estimate any of the variance terms in a model, let alone accommodate the sophisticated partitioning of uncertainties provided by hierarchical models. One general approach that may prove fruitful, when it is computationally feasible, would be to use a full state-space model, fit by MCMC, when calibrating a model and its error parameters against past observations, and for making the initial forecast. Indeed, for simple models with small data sets the process of updating the forecast may be done by refitting the full state-space model each time new observations become available. If a full refit is not possible, one might choose to update the forecast from the full state space by treating the MCMC output (both the forecast and the parameter estimates) as informed prior samples for a particle filter. Alternatively, computational cost might be further reduced, at the cost of more stringent assumptions, by instead generalizing the ensemble-based approach and approximating these priors with analytical distributions.

To give an example on how data assimilation can be generalized, figure 14.4 illustrates an alternative to the most basic Kalman filter (univariate, linear, Gaussian), where the process variance, $q$, observation variance, $r$, and the model slope, $m$, are treated as unknown fit parameters rather than assuming they are known. Both $q$ and $r$ were reparameterized as precisions with uninformative Gamma priors, while $m$ and the initial condition of $X$ were given uninformative Normal priors. The model was fit sequentially, one data point at a time, using a Monte Carlo approach for the forecast step. However, rather than doing the analysis step as a particle filter, after each forecast the analytical, named prior distributions were updated based on moment-matching, and then for the analysis both the state and the parameters were updated using MCMC. In other words, the model is fit using a generalized ensemble-based approach, where all parameters are updated in the analysis, rather than using the EnKF.

The whole process of fitting the model was repeated under three different scenarios to illustrate the impacts of relaxing different assumptions. I started with a model where only the process variance was treated as unknown, since I have argued elsewhere in the book that this is a critical component to ecological forecasting and is difficult to estimate independently. Under this scenario, the model converges rapidly and gives results very similar to the KF or EnKF (not shown). Next, both the observation and process variances were treated as unknowns. In this case the forecasts were uninformative for the first ~7 observations. After this the posterior estimates began to converge and by ~15 observations estimates were similar to the previous scenario, albeit with slightly greater uncertainty. Because uninformative priors were used for both $q$ and $r$, the model had difficulty distinguishing the two, which led to an overestimation of observation error and an underestimation of process error. The notable difference between these scenarios reiterates the critical importance of including informative estimates of observation uncertainty with data (chapters 3 and 6). Finally, adding process model parameter error in the third scenario generates a pattern similar to the second scenario, but with slower convergence because additional unknowns need to be estimated. That said, once the slope estimate converges; the second and third scenarios are indistinguishable.

Overall, within the general concept of the forecast cycle any of the techniques introduced for uncertainty propagation (forecast) and partitioning (analysis) can be used to construct new data assimilation filters that accommodate the assumptions you are willing to make. In general, stronger assumptions result in greater computational efficiency, so the larger the problem (bigger data or slower models), the greater the trade-off.

## 14.5 KEY CONCEPTS

1. Ensemble filters, such as the Ensemble Kalman filter (EnKF), apply ensemble methods to propagate uncertainty in the forecast step. They flexibly capture nonlinearities, can capture multiple sources of uncertainty, and are relatively easy to implement because they don't require a Jacobian, but come at the cost of requiring more model runs.

2. Particle filters apply a full Monte Carlo approach in the forecast step. The analysis step is performed by weighting the different ensemble members based on their likelihood.

3. Model averaging and reversible jump MCMC provide means for forecasting with multiple process models simultaneously, thus accommodating model structural error.

4. Sequential data assimilation methods can easily be generalized, with any uncertainty propagation technique used in the forecast and any Bayesian computation (for example, MCMC) used in the analysis.

5. Existing data assimilation methods are aimed specifically at dealing with the Initial Condition problem. An open question is, what simplifying assumptions to the general state-space framework are appropriate for ecological forecasting?

## 14.6 HANDS-ON ACTIVITIES

https://github.com/EcoForecast/EF_Activities/blob/master/Exercise_10_ParticleFilter .Rmd

- Particle filter

# 15

## Epidemiology

SYNOPSIS: *Understanding the progression of contagious diseases in humans and other organisms is of great public interest, with forecasts focused on predicting the start of an outbreak, the timing of the peak, and the impact of interventions. This forecasting is challenged by the chaotic nature of outbreaks, their sensitivity to stochastic events, and high parameter uncertainty due to rapid evolution. Epidemiological forecasting has to occur in real time with limited (and sometimes biased) data to make rapid decisions. This chapter's case study on H1N1 highlights the use of sequential data assimilation (chapters 13 and 14) to update outbreak forecasts in real time.*

WHETHER FROM NEWS reports on antibiotic resistance, terrifying stories about Ebola, epidemic thriller movies such as *Outbreak* and *Contagion*, or fictional viruses leading to global apocalypse (or vampires and zombies), the fear of infectious disease is deeply seated in the human psyche. This fear is not unfounded, as infectious disease is responsible for over a quarter of all human deaths annually (WHO 2004). Historical pandemics of bubonic plague, influenza, and smallpox killed tens of millions of people per outbreak and occurred as recently as the 1918 Spanish flu. Endemic tuberculosis killed one-quarter of the population in nineteenth-century Europe. Diseases introduced to the New World by European explorers, such as smallpox, may have had mortality rates of 80% to 90% among native peoples who lacked prior exposure (Diamond 1999). In addition to historical scourges, there is growing concern about emerging infectious diseases as a result of population growth, globalization, and climate change. In 2011 there were 34 million people with HIV and 1.7 million deaths (WHO 2013). The 2002 SARS outbreak and 2009 H1N1 Influenza outbreak both resulted in global impacts on society and the economy, though thankfully neither was as deadly as initially predicted.

In addition to the direct impacts on humans, diseases in animals and plants have substantial ecological, biogeochemical, economic, and social impacts. For example, the 2001 UK outbreak of foot-and-mouth alone resulted in 4 million livestock being culled and had a £6 billion ($9.3 billion) economic impact on the UK (Thompson et al. 2002). The Great Irish Famine in the 1840s that killed over 1 million people was triggered by a potato blight (*Phytophthora infestans*). To combat crop pests and diseases, modern industrial agriculture applies approximately 5.2 billion pounds of pesticides per year at a cost of almost $40 billion/year (Grube et al. 2011). This pesticide usage leads to a whole cascade of human health and environmental impacts, causing an additional $10 billion in damages in the United States alone (Pimentel 2005). For native plant populations, invasive plant diseases such as chestnut blight (*Cryphonectria parasitica*) and white pine blister rust (*Cronartium ribicola*) have resulted in the functional elimination of keystone species, while endemic infections have been estimated to reduce forest productivity by ~6% in the United States and ~16% in Canada (Hatala et al. 2011; Hicke et al. 2012). Indeed, there are few areas in pure or applied ecology where the risk of disease doesn't impact the dynamics of the systems being studied.

The following sections will (1) briefly summarize the theory behind infectious disease models and the challenges of disease forecasting; (2) highlight examples of research on foot-and-mouth and measles; and (3) explore the details of a case study on H1N1 influenza forecasting.

## 15.1 THEORY

The primary forecasting problem in epidemiology is to predict disease outbreaks in populations. This area has benefited from a good bit of attention, as the problem is fundamentally the same whether it is applied to humans, animals, plants, or microorganisms. Unlike many other areas in ecology, simple models have been remarkably useful for capturing disease dynamics. Known as compartment models in human epidemiology, these models primarily come from the long-standing Susceptible-Infected-Removed (SIR) family (Kermack and McKendrick 1927).

The basic SIR model (figure 15.1) assumes a fixed population size, $N = S + I + R$, with individuals transitioning among groups as follows:

$$\frac{dS}{dt} = -\beta SI$$

$$\frac{dI}{dt} = \beta SI - \gamma I$$

$$\frac{dR}{dt} = \gamma I$$

The contact rate, $\beta$, is the rate at which an infected individual (I) will encounter and infect a susceptible individual (S), leading to an overall flux from susceptible to infected at the rate $\beta SI$. The flux of infected individuals to the removed/recovered pool (R), is controlled by $\gamma$, the recovery rate. The recovery rate is often more intuitively understood as $1/\gamma$, the average infectious period. There are other variants in the SIR family of models that add a stage where individuals are exposed (E) but not infectious (SEIR), when individuals have no immunity after infection (SIS) or lose immu-

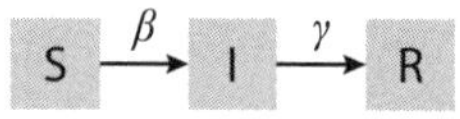

$$S \xrightarrow{\beta} I \xrightarrow{\gamma} R$$

FIGURE 15.1. Structure of the Susceptible-Infected-Removed (SIR) model. This model represents the number of individuals in each stage, with transitions controlled by the contact rate, $\beta$, between susceptible and infected individuals and the recovery rate of infected individuals, $\gamma$.

nity after some period (SIRS), or when infants are conferred with temporary maternal (M) resistance (MSIR, MSEIR), and so on.

For a disease outbreak to increase requires that $dI/dt > 0$. Solving for this condition gives a simple prediction for a threshold in the *basic reproductive number*, $R_0 = \beta S_0/\gamma$, between an increasing ($R_0 > 1$) versus declining ($R_0 < 1$) epidemic. The initial pool of susceptibles, $S_0$, is often the size of the population, $N$, but can often be lower if past infection or immunization creates an initial pool of removed individuals. For a given set of population parameters, this model predicts a threshold population size, $S_0^* = \gamma/\beta$, above which a disease will increase in abundance if an infected individual is introduced into a population.

We also learn that the risk of outbreak increases with both average length of infection and contact rate. From a management perspective there is little we can do about the infectious period, but it is possible to reduce the contact rate, for example, through a quarantine, or reduce the effective population size through immunization. This result further predicts that a population is protected from epidemic so long as the number of unvaccinated individuals stays below $S_0^*$, a concept known as herd immunity.

## 15.2 ECOLOGICAL FORECASTING

The simplicity of these critical thresholds in epidemiology has been a mixed blessing for disease forecasting. On the positive side, predictions can be made by estimating two parameters, which doesn't require simulating the dynamics or more advanced data assimilation techniques. On the negative side, these simple solutions have resulted in a strong focus on analytical thresholds, which has made it easy to ignore the effects of uncertainties (for example, model/process error, uncertainties in parameters and initial conditions) on outbreak dynamics.

As with other areas of ecological forecasting, we can separate the forecasting problem into predictions based on the current state versus projections under different future scenarios. Over short timescales the classic forecasting problem is to determine the risk that an outbreak will occur, the size of the outbreak, and when it will peak and decline. Over longer timescales the disease forecasting problem often shifts to the invasive species problem of anticipating spatial spread. At both timescales there is often an urgent interest in projections evaluating disease intervention scenarios.

Epidemiological forecasts are typically initiated for populations with small numbers of infected individuals, and there is often considerable uncertainty about how many individuals have already been infected. At these low levels chance events can play a large role in whether an infection dies out or not, and in determining how long it takes to get going. That said, if $R_0 > 1$, then an emerging infection is inherently unstable (chapter 2) and can display the high sensitivity to initial conditions indicative of chaotic dynamics. Finally, initial forecasts also typically start with poorly constrained estimates of the model's parameters, since there have been few observations from which to estimate vital rates. Because of these four factors (sensitivity to initial conditions, stochasticity, uncertainty in initial conditions, and uncertainty in parameters) disease forecasting remains challenging and perpetually data-limited. The first three of these factors can have enormous impacts on the occurrence and timing of outbreaks and illustrate why quantifying $R_0$ by itself is not sufficient. They also suggest why disease forecasting may be productively viewed as a state-variable data assimilation problem, because even with perfect knowledge of model parameters there will often be considerable forecast uncertainty and the trajectory needs to be continually updated as new information becomes available. However, unlike most other chaotic systems we've discussed, once an outbreak is under way the model projections exhibit a decreasing sensitivity to stochasticity (figure 15.2), and indeed at the peak of an outbreak the system switches over to a period of predictability (negative Lyapunov exponents; see box 2.2) that extends through the decline and into the quiescent period between outbreaks (Grenfell et al. 2002). As the pool of susceptible individuals is replenished, the system once again becomes increasingly unstable and sensitive to stochastic events.

## ▨▨▨ 15.3 EXAMPLES OF EPIDEMIOLOGICAL FORECASTING

While there is a long history of predictive SIR modeling in both human epidemiology and ecology (Anderson and May 1983), the 2001 foot-and-mouth disease (*Aphthae epizooticae*, FMD) outbreak in the UK is viewed by many as an important turning point in ecological forecasting. FMD is a highly infectious virus to which a wide range of cloven-footed mammals, including livestock such as cattle, sheep, and pigs, are susceptible. Cases were first reported among pigs in the northeast of England on February 19, 2001, after a 34-year period without a major outbreak, and subsequently spread rapidly among sheep and cattle by a combination of local contact, windborne dispersal, and the long-distance transport of livestock (Ferguson et al. 2001a). The imposition of movement restrictions in late February was estimated to have reduced $R_0$ from 4.5 to 1.6, which was substantial but insufficient to bring $R_0$ below 1 and halt the epidemic. An SEIR-type model, modified to account for spatial structure, was fit to available data and on March 28, 2011, predicted 44% to 64% of the population was at risk of infection (Ferguson et al. 2001a). The model was then used to evaluate a range of scenarios, such as on-farm slaughter, vaccination, and ring-culling, the preventative slaughter of all animals within a certain radius of each new diagnosis. Based on these analyses, a ring-culling government policy was introduced, which was shown to have been effective in controlling the epidemic (Ferguson et al. 2001b). This result was largely corroborated by further farm-level modeling assessing dispersal and the high degree of spatial heterogeneity (Keeling et al. 2001b), though neither conclusion was free of critics arguing that the ring-culling was excessive (Kitching et al. 2005). The near real-time predictions of the spread of FMD, coupled with the scenario-based evaluation of management options, have been credited not only with helping to control this specific outbreak but also with helping disease modeling to become a more predictive science (Tildesley et al. 2008). The successful control of the 2001 outbreak based on model *predictions* led to further *projections* exploring strategies for both prophylactic and reactive vaccination as means for controlling future outbreaks (Keeling et al. 2003; Tildesley et al. 2006).

At roughly the same time that ecologists in the UK were providing real-time forecasts of FMD, another team was investigating the historical dynamics of UK measles outbreaks prior to the vaccination era. Building upon previous work on the roles of noise, nonlinearity, and seasonal forcing on disease dynamics (Ellner et al. 1998; Keeling et al. 2001a), they introduced a stochastic, time-series version of the SIR model—a model referred to as the TSIR (Bjørnstad et al. 2002; Grenfell et al. 2002). The TSIR approach is non-Gaussian, nonlinear, discrete in both time and individuals,

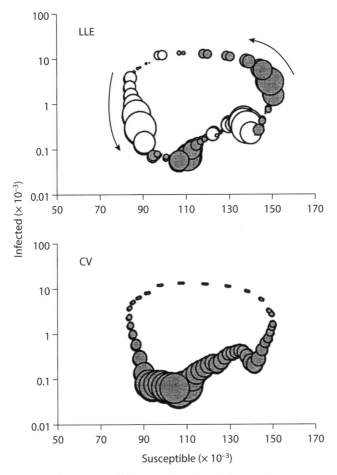

FIGURE 15.2. Impact of system stability and stochasticity on the two-year measles epidemic cycle in London (1944–65). Arrows depict the counterclockwise direction of the two-year cycle (figure 15.3) for the rising and falling legs of the larger outbreak year, with the outbreak rising from the 3 o'clock position to peak at the 12 o'clock position (maximum number of infected individuals) and then declining to a minimum at the 6 o'clock position. A second, lower peak occurs in alternate years around the 4 o'clock position and is followed by a more modest decline. The size of each circle is proportional to the magnitude of the local Lyapunov exponents (LLE, top), an index of system stability, or the one-step-ahead coefficient of variation (CV, bottom), a measure of the impact of stochasticity. Open circles in the top figure illustrate periods of the disease cycle where the feedbacks are stabilizing (negative LLE), and thus the dynamics are more predictable, while the gray circles indicate periods when the dynamics are unstable (chaotic, positive LLE). Generally, dynamics are most unstable at the start of an outbreak and most stable during the decline. The bottom figure illustrates that the effects of stochasticity are largest when the number of infected individuals is low. This combination of instability and stochasticity at the start of an outbreak makes forecasting the timing and magnitude of outbreaks particularly challenging. Figure adapted from Grenfell et al. (2002).

and allowed for seasonal variability in contact rate. Unlike the traditional approach to ecological theory, which focused on the deterministic core of model processes and dynamics, the TSIR model is very much a *statistical model*. It was designed to be fit to data and capture multiple sources of uncertainty and stochasticity. Specifically, the TSIR model assumes that the number of new infections at the next time step is distributed as a Negative Binomial (NB) distribution

$$I_{t+1} = NB(\lambda_{t+1}, I_t)$$

where the expected number of new infections, $\lambda_{t+1}$

$$\lambda_{t+1} = \beta_t (I_t + \theta_t)^{\alpha} S_t$$

is based on a modified form of the standard SIR that accounts for immigration of infected individuals from other areas, $\theta_t \sim \text{Pois}(m)$, where $m$ is the immigration intensity. The parameter $\alpha$ allows for spatial structure in the mixing among individuals, relaxing the traditional "mass action" assumption that susceptible and infected individuals encounter each other at random. The Negative Binomial (table 6.1; figure 6.1) is a discrete distribution that describes the number of trials needed to produce the specified number of events (for example, how many coins with probability $p$ do you have to flip to get $n$ heads). An alternative derivation of the NB is as a Poisson process with a Gamma distributed mean, thus producing a distribution similar to the Poisson but with greater variability. In addition, to account for the seasonality in measles epidemics the contact rate, $\beta_t$, is modeled as depending upon the time of year (26 different values, one for each two-week period within a year). Finally, since measles is predominantly a disease among young children, the susceptible pool was modeled as

$$S_{t+1} = S_t + B_t - I_t$$

where $B_t$ is the time-varying number of births into the population, which is estimated from public records.

The novelty of the TSIR approach wasn't just the model itself, but the depth of the analysis that it enabled. The model was fit to biweekly data from 60 British cities from 1944–65. The balance between noise and determinism in controlling outbreaks was shown to be a function of population size, with lower population cities being more subject to demographic stochasticity, local extinction, and reintroduction (Grenfell et al. 2002). By contrast, large cities were much more stable in their cycles, with switches in periodicity being driven by changes in birth rate rather than complex dynamics (figure 15.3). At a regional scale, these effects drive a core-satellite metapopulation, with larger cities serving as the "cores" and synchronizing the outbreaks in surrounding "satellite" areas at a slight lag to that of the cities (Grenfell et al. 2001). Furthermore, the timing of the periodic outbreaks was strongly regulated by the seasonal variability in contact rate, which was higher when school was in session and lower during school holidays (Bjørnstad et al. 2002).

In the wake of the TSIR framework, forecasting research on measles has extended out to parts of the world where measles epidemics have occurred more recently, such as Bolivia, Cambodia, and Pakistan, and others, such as Niger and Nigeria, where epidemics are unfortunately still common (Ferrari et al. 2008; Bharti et al. 2011; Chen et al. 2012). This work has demonstrated a range of dynamics more complex than that exhibited in the UK, where the interactions between population growth and

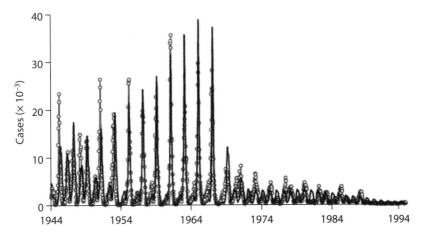

FIGURE 15.3. London measles outbreaks from 1944–94. Gray lines and circles are observations, while the black line depicts model predictions. The TSIR model was calibrated to the prevaccination period (1944–65), and predictions for 1966–94 were made adjusting effective birth rates for the level of vaccination (that is, vaccinated individuals were not introduced into the susceptible pool). The annual cycle prior to ~1950 is driven by the post–World War II baby boom, while the remainder of the prevaccine period exhibited the two-year cycle shown in figure 15.2. Figure adapted from Grenfell et al. (2002).

seasonality can produce unexpected results, such as a shift from periodic to chaotic dynamics in response to vaccination (Ferrari et al. 2008). More recent work also tackles the forecasting problem more explicitly (Chen et al. 2012) and has demonstrated the usefulness of data fusion, for example by utilizing remote sensing to track shifts in population density (Bharti et al. 2011). Beyond measles and foot-and-mouth, there's been excellent work done fusing models and data to understand emerging infectious diseases such as SARS, Dengue, Lyme, and West Nile (LaDeau et al. 2011). Progress has also been made in predicting plant, animal, and zoonotic disease at the landscape scale (Meentemeyer et al. 2012), though much work remains to be done to connect disease in natural systems to ecosystem processes (Hicke et al. 2012; Dietze and Matthes 2014).

## 15.4 CASE STUDY: INFLUENZA

In this case study I use influenza forecasting to highlight the real-world application of sequential data assimilation (chapters 13 and 14) to ecological forecasting. Compared to many other diseases that have been greatly reduced or eliminated by vaccination campaigns, the high contagion and rapid evolution of the common flu leaves it remarkably difficult to either control or predict. Indeed, there have been a wide range of data assimilation techniques applied to influenza forecasting, a trend that appears to be increasing as flu infection estimates are increasingly available in near-real time (see review in Niu et al. 2014). In addition, influenza pandemics have occurred in modern times (for example, H1N1 in 2009), and the threat of future pandemics remains real. While the H1N1 epidemic of 2009 was not as severe as initially feared, it still had significant human health and economic impacts, and such threats need to be taken seriously.

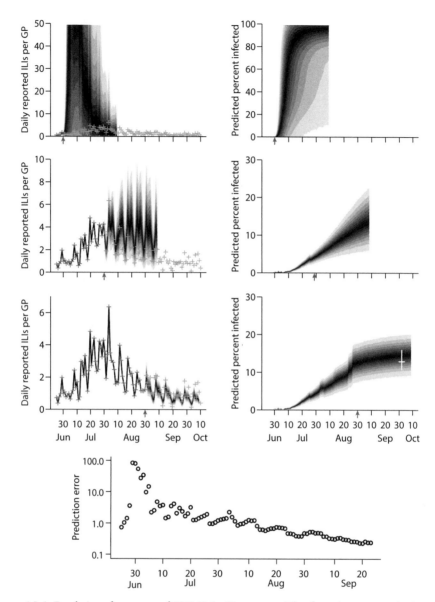

FIGURE 15.4. Real-time forecasts of H1N1 in Singapore. The first three rows depict observations (crosses) and forecasts (gray shaded areas) at three points in time during the outbreak (indicated by the arrow on the time axis). The cross in the lower right indicates an independent estimate of infection from adult seroconversion. The bottom panel shows the posterior absolute deviation between predicted and observed incidence. Figure reproduced from Ong et al. (2010).

On a more local scale, the 2009 H1N1 outbreak also provides an excellent case study on the potential for real-time ecological forecasting to impact human decision making, on both the personal and policy levels (chapter 17). As the H1N1 outbreak started to get under way, a grass-roots group of researchers at local hospitals and universities in Singapore pulled together a monitoring effort and forecast system for the city over the course of a few weeks and with no budget (Ong et al. 2010). These researchers solicited local clinics to e-mail or fax in daily reports of influenza-like illness. The data was used not only to track disease progress but also to generate a *daily update* to a probabilistic forecast using an SEIR model. Specifically, they implemented a discrete-time, discrete-individual version of the SEIR where the numbers of individuals undergoing each transition was assumed to follow a binomial distribution. Furthermore, in addition to estimating the state variables the model also updated the model parameters each day, allowing initially rough estimates to become iteratively more constrained in light of new observations. Observations were only available on the number of new infections, which was assumed to have Poisson observation error, with the other three states treated as latent variables (chapter 9). This data model was further adjusted for the day of week, as new reported cases tended to be lower over the weekend and highest on Monday, and for the fraction of the local population served by the clinics that participated in the study. This model was fit as a Bayesian state-space model (chapter 8), and the forecast was *updated every day using a particle filter* (chapter 14). Unlike most earlier disease forecasting papers, this model was implemented as a genuine "operational" data assimilation workflow (chapters 3 and 4), run sequentially as new observations became available each day, and pushed to a public website that received local press attention. During the early phases of the infection, when the forecasts were based primarily on the semi-informative priors, the model tended to overpredict infection, both in terms of the proportion of the population that would eventually be infected and the speed of the epidemic (figure 15.4). As the epidemic progressed the prediction error declined. The model successfully predicted the peak infection approximately two weeks in advance and then captured the decline phase, and the final proportion of the population infected (~13%), with high fidelity.

Overall, disease epidemiology faces challenges, such as high parameter uncertainty and high sensitivity to initial conditions and stochastic events, that were present in all our case studies but particularly exacerbated in this domain. However, because of this and the high social relevance of disease, epidemiology has been at the forefront of near-real-time forecasting. Disease ecologists also have an established track record for conveying this information to the public and providing decision support to policymakers (chapter 17).

## 15.5 KEY CONCEPTS

1. Simple SIR-family models have had success in disease forecasting but have overly emphasized deterministic thresholds instead of dynamics.
2. Disease forecasting remains challenging due to sensitivity to initial conditions, stochasticity (process error), uncertainty in initial conditions, and uncertainty in parameters.
3. Disease forecasting provides examples of operational forecasts that combine data assimilation and informatic workflows to inform decision making.

# 16

## Assessing Model Performance

*Synopsis: One of the reasons weather forecasts have improved so much over the decades is that every day they make specific, quantitative predictions that can be tested against observations. While most of this book has focused on uncertainties that can be captured by probability distributions (parameters, initial conditions, drivers, process error), model assessment is one of the main ways we assess errors in model structure, and thus provides an important feedback loop to improving not just the models themselves, but also the underlying theories and assumptions upon which models are built. Furthermore, model assessment is critical for identifying the conditions under which a model is reliable versus where it breaks down—just like Newtonian physics, a model can be "wrong" but still be useful under the right conditions. Here we discuss a diverse array of techniques that help us assess the skill of our forecasts and identify when and where they fail.*

### 16.1 VISUALIZATION

The first stage of model assessment is a "sanity check." Calculating basic statistics on model outputs (mean, standard deviation, quantiles) and plotting model predictions ensures that the model is qualitatively behaving as expected and outputting in the expected units. There is no point in going through the effort of checking models against data if it is already obvious that the model is very wrong (that is, producing results that are not just a little off, but completely implausible). A lot of the failures at this stage will result from bugs in the code or improper units on calculations, while at other times these checks may identify false assumptions in the model construction.

The second stage of model assessment involves a graphical comparison of models to data to assess performance. One of the most basic and routine checks is a *predicted/observed plot*—a scatter plot of observations (on the $y$-axis) versus predictions (on the $x$-axis). Ideally, these predictions should fall on a 1:1 line. This plot allows for the detection of outliers and the qualitative identification of bias or model miscalibration (for example, differences in intercept or slope from the 1:1 line). When possible, it is also useful to include error estimates on both the observations and predictions when making predicted/observed plots, as this helps in judging which errors are significant. For example, if you put a 95% interval on a model prediction, 95% of these predictions should overlap the 1:1 line. Large scatter in this plot can be indicative of either low skill, noisy data, or both, depending on whether the uncertainties are in the $X$ or $Y$ direction. By contrast, consistent deviations from the 1:1 line indicate systematic errors and can help identify conditions under which the model is failing. Scattered, individual outliers can present a more challenging problem in diagnosing the conditions that led to model failure. Overall, when either outliers or systematic errors are observed in a predicted/observed plot, it is generally safe to start from the assumption that these are errors in the model. That said, it is always important to understand the data you are using and how it was collected, as sometimes the measurements themselves are the source of both systematic error and outliers.

After a basic predicted/observed plot, it is common to *plot model projections in time or space to assess the modeled dynamics*. For example, compared to a miscalibration, which is often easy to assess in the predicted/observed plot, a shift in the timing of events may lead to errors that are visually easy to assess in a time-series plot but hard to detect in a predicted/observed plot or using basic diagnostic statistics.

Beyond these basics, it is hard to underestimate the value of *creative diagnostic visualizations*. Do models capture higher-level patterns and emergent phenomena? Do model outputs capture the relationships among output variables and between inputs and outputs? For example, Schaefer et al. (2012) plotted the relationship between GPP versus light and humidity, two of the main drivers of photosynthesis, for both process-based terrestrial ecosystem models and statistical models based on flux-tower observations (figure 16.1). As a function of light, most terrestrial ecosystem models captured the shape of the response, but many models underestimated GPP. This suggests models were capturing the process but had the wrong parameters, and thus most models should be easily improved by better calibration against data (chapter 5). By contrast, for a number of models the effects of humidity on GPP were qualitatively different in shape from observations. This is not the type of pattern that is corrected by data assimilation, but instead represents an error in model structure. Furthermore, while it doesn't point to a specific mechanism of failure, it does identify specific conditions associated with failure, which is a critical first step toward identifying the underlying processes causing the error and testing alternative hypotheses and assumptions.

Overall, diagnosing model performance is very much analogous to *hypothesis testing*. Based on the observed patterns in the model, and comparisons to data, you begin to formulate hypotheses about where, when, and why a model is failing. From this hypothesis, you may propose some change to the model (that is, an experiment) and make a prediction about how that change will affect model output. Experiments might involve changes to model parameters (for example, sensitivity and uncertainty analyses, chapter 11) or changes to model structure (for example, the implementation of some new function or subroutine). If there are competing hypotheses about the observed pattern, *computational experiments* should be designed that lead to different predictions. Finally, you would generate runs from the model to evaluate these computational experiments, which would then support or refute the prediction. One of the great things about models is that you can generally perform many more experiments, under a much wider range of conditions (some of which may be impossible in the real world), than you could ever do in the physical world.

Applying this hypothesis-testing perspective to the previous GPP example, for the light response our hypothesis was that the model was miscalibrated. Specifically, we might predict that the parameters controlling the maximum photosynthetic rate might

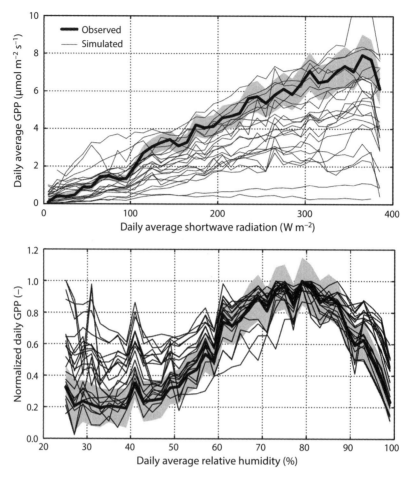

FIGURE 16.1. Diagnosing errors in GPP functional responses. Each gray line is the simulated response from a different ecosystem model, while the gray envelope around the black line represents the data uncertainty. The top figure suggests most models capture the shape of the light functional response, but many underestimate the slope. By contrast, the bottom figure suggests models capture the functional response at high humidity, but many diverge at low humidity (Schaefer et al. 2012).

be too low. Alternatively we might hypothesize that these models had a leaf area index (LAI) that was too low (that is, not enough leaves). In both cases we predict that increasing these parameters will increase GPP. These alternatives could be assessed by performing a sensitivity or uncertainty analysis, varying the parameters in consideration over a prior range consistent with their uncertainties. This prior restriction is imposed because a photosynthetic rate or LAI that "solves" the problem using biologically implausible parameter values would result in getting the right answer for the wrong reason. If only one of these analyses brings you within the range of observations, it would support that hypothesis. If both did, it may suggest a need for further field measurements, while if neither is sufficient, then we have to go back to the drawing board and explore additional hypotheses.

## 16.2 BASIC MODEL DIAGNOSTICS

After the performance of a model has been qualitatively assessed, it is frequently useful to perform quantitative assessments. The goal of quantitatively assessing model skill is not to generate a pile of error statistics that you then put in a table that no one reads, nor is it to pat yourself on the back and demonstrate what a good job you did at modeling. Instead, the goal within a single model is usually to quantify and summarize the patterns identified in the visual analysis and thus dive deeper into the identification of *structural errors* in the model. Likewise, when looking across models (whether wholly separate models, or different versions of a single model), the goal is not to rank models as good or bad but to assess the relative performance of different model structures. Remember, at their root models are just a distillation of our current working hypotheses, and thus comparing alternative models is at its heart aimed at testing alternative hypotheses about how the world works. Finally, since our visual systems are exceedingly good at finding pattern, even when it's not actually there, quantitative statistics can also help prevent us from over interpreting visual diagnostics.

Common model diagnostics include root mean square error (RMSE), correlation coefficient, $R^2$, regression slope, and bias (Hyndman and Koehler 2006). Among these, RMSE is probably the most important metric, as it is essentially the standard deviation of the distribution of residuals (figure 16.2, top):

$$RMSE = \sqrt{\frac{1}{n}(Y_{model} - Y_{data})^2}$$

As such it is the single most common summary of model error and the most important metric for propagating model error into further analyses.

The *correlation coefficient* gives a measure of whether the model and observations are related to each other (figure 16.2, middle), but not whether the model is well calibrated—one could have a correlation of 1 and still not fall on the 1:1 line.

In a regression model, the $R^2$ is just the square of the correlation and doesn't provide any additional information, but in the context of model diagnostics it is *more appropriate to report the $R^2$ in terms of the deviations from the 1:1 line*

$$R^2 = 1 - \frac{\sum (Y_{model} - Y_{data})^2}{\sum (Y_{data} - \overline{Y_{data}})^2}$$

rather than the deviations from the regression between predictions and observations. The latter is unfortunately prevalent but is misleading, as it can create the false impression that a miscalibrated or biased model is explaining a large fraction of the observed variability. The *slope of the regression* between model and observations can serve as a metric of miscalibration, and ideally should be 1 (as should the correlation and $R^2$). If any statistical tests are performed on this slope, it is critical that the appropriate null model is chosen (slope = 1), as this is *not* the default hypothesis in statistical packages (slope = 0), and therefore any F-tests, t-tests, and $p$-values from that default null are not the ones you are looking for.

*Bias* indicates whether a model is consistently under- or overpredicting and is calculated simply as the difference between the modeled and observed means. Similarly, the *ratio of the standard deviation* in the model output to the standard deviation in the data, $SD[Y_{model}]/SD[Y_{data}]$, is useful for determining whether the model is

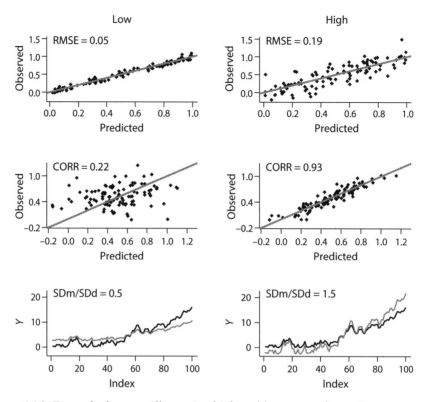

FIGURE 16.2. Example data sets illustrating high and low scores for various statistics. Top row: Root mean square error (RMSE). Middle row: Correlation coefficient. Bottom row: Plot of model and data time-series (gray = model, black = data) illustrating differences in the ratio of standard deviations (model/data).

under- or overpredicting the amount of variability observed in the real world. In particular, it helps diagnose whether the model is smoothing the dynamics too much (model variance < observed variance) or predicting fluctuations that are not observed (model variance > observed variance). For example, in figure 16.2, bottom, these two time series would have very similar RMSE, bias, and correlations, but the right-hand plot is overpredicting the true variability while the left is underpredicting the variability.

Among all the metrics mentioned earlier, one thing that is noteworthy is that none account for the uncertainties in either the model or the data. The *reduced chi-square* metric is similar to the $R^2$ but normalizes by the observation error rather than the distribution of observations

$$\chi^2_{red} = \frac{1}{N-1} \sum \frac{(Y_{model} - Y_{data})^2}{\sigma^2_{obs}}$$

A reduced chi-square greater than 1 indicates that the model-data residuals are more variable than expected based on the error in the observations alone (that is, the error in the model is greater than the error in the data). Similarly, a common approach for accounting for model uncertainty is to calculate the predicted quantile of each ob-

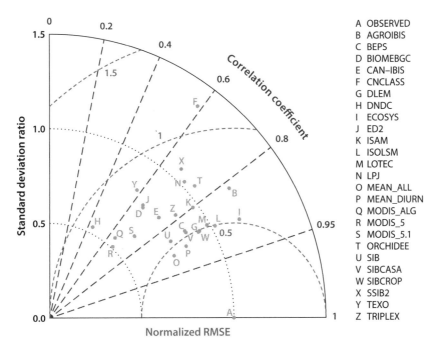

FIGURE 16.3. Taylor diagram assessing the ability of 20 ecosystem models to predict daily mean GPP across 39 eddy-covariance towers. Data are located at point A. The multi-model ensemble mean (O, P) had the lowest RMSE and highest correlation but underpredicted the true variability. A cluster of models (G, M, L, W) had a slightly lower correlation and higher RMSE but had similar variability to the observations. The MODIS GPP products (Q, R, S) performed worse than most models and substantially underpredicted the true variability in the system (Schaefer et al. 2012).

served data point—sometimes referred to as a *Bayesian* p-*value*—rather than simply subtracting the difference between the model and data. A histogram of Bayesian *p*-values should have a uniform distribution. Similarly, most of the visualizations and statistics discussed earlier can be performed using these quantile scores. In addition, all of the earlier metrics focused on continuous response variables. When making probabilistic forecasts of categorical variables (for example, whether a predicted event occurs), the *Brier score* is a useful metric for assessing how well the forecast probabilities, *f*, match the binary outcomes, *o*

$$BS = \frac{1}{N}\sum_{t=1}^{N} \sum_{i=1}^{R} (f_{it} - o_{it})^2$$

where $N$ is the number of forecasts being assessed and $R$ is the number of possible outcomes (for example, $R = 2$ for a Boolean variable, such as rain/no rain). Low scores are better (perfect = 0, perfectly wrong = R). Relative to just getting the mean background rate correct (that is, guessing $f_{it} = \bar{o}$ on every forecast), this metric rewards more confident predictions but strongly penalizes overconfident wrong predictions.

The *Taylor diagram* (figure 16.3) is a particularly handy visual tool for summarizing and comparing models as it combines the three error statistics illustrated in figure 16.2: RMSE, correlation, and the ratio of model variability to observed variability.

When reading a Taylor diagram, the observed data are always located on the $x$-axis at (1,0) and the "best" model is the one that is closest to the data. In plotting models on the Taylor diagram, the correlation coefficient is plotted as rays emanating out from (0,0), starting with a correlation of 1 plotted along the $x$-axis and rotating counterclockwise until reaching a correlation of 0 along the $y$-axis. In the event that models are negatively correlated with observations, the rays are extended into the left-hand quadrant, with a correlation of –1 occurring along the negative $x$-axis. The standard deviation ratio is plotted as distance along the correlation ray, with reference arcs, such as $SD_{model}/SD_{data} = 1$, frequently being plotted on the diagram as well. Essentially, *the Taylor diagram is a polar plot of these two pieces of information (standard deviation ratio and correlation)*. What makes the Taylor diagram even more informative is that, when plotted this way, the RMSE of the model ends up being represented as distance from the observations—the model with the lowest error is closest to the point representing the data. Because of this property, Taylor diagrams are especially useful for comparing among multiple models, comparing the performance of a model at multiple sites or conditions, and assessing the change in model performance over time as model development occurs. For the last case it can be additionally helpful to plot vectors between different iterations of a model to indicate the direction of model error over time.

In assessing model performance it is always more powerful to check the model against independent data from that which was used to fit the model. If the model predicts multiple outputs, it is likewise very helpful to check a model against more than one output, even if these come from the same site as was used for model calibration. It is easy for models to be optimized to predict one variable well at the expense of giving poor, or even biologically implausible, predictions for another variable (chapter 9), which clearly indicates an error in model structure. In the event that independent data are unavailable, it is common practice to hold out some portion of the original data during calibration for the purpose of verification. A related approach, known as *cross-validation*, involves *repeatedly holding out a subset of data, fitting the model to the remaining data, and then predicting the held out portion*. This tries to balance the advantages of having an independent validation (for example, avoiding overfitting) while still using all the data in hand. How the data are partitioned into the fit versus reserved portions can have important impacts on the inference made during cross-validation. In general, cross-validation is more robust if the withheld portions are chosen in a stratified manner (that is, grouped systematically according to one or more covariates), rather than randomly (Wenger and Olden 2012). For example, if observations exist on a latitudinal gradient, one might leave out whole latitudinal bands, forcing the model to extrapolate across the region. Likewise, leaving out larger portions of data tends to be more robust than smaller subsets (with the limit being leave-one-out cross-validation, which uses $n - 1$ observations to predict the missing one).

In addition to comparing model outputs to observations of the same variable, a very useful approach to model checking is to verify that the model correctly predicts observed *relationships among output variables, and between inputs and outputs*. For example, as we did with the earlier GPP example, one could check that a model correctly predicts the observed relationship between some environmental driver (for example, light) and the observed ecosystem response (figure 16.1). This is especially useful when such relationships are emergent phenomena in the model rather than

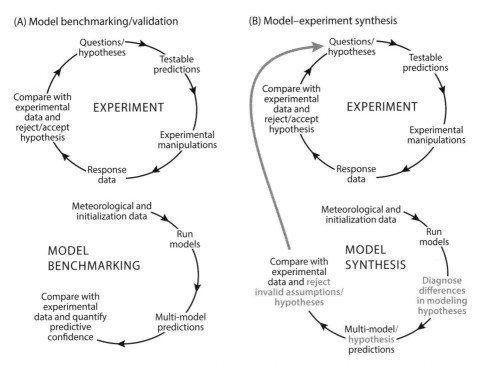

FIGURE 16.4. Contrast between model-data benchmarking and model-data-driven hypothesis testing (Walker et al. 2014).

prescribed functional forms, but are important checks even in the latter case. As mentioned earlier, a key aspect of model diagnosis is to approach the problem from a hypothesis testing framework (figure 16.4). Whether evaluating a single model, or comparing multiple alternative models, the process of breaking down complex model outputs into specific hypotheses can be a powerful tool in assessing whether models are not just generating the correct outputs, but are doing so for the correct reasons (Medlyn et al. 2015). Overall, it is important to *focus on the key assumptions of models* rather than just plugging through routine diagnostics. It can frequently take some creativity, and always requires insight into how a system functions (rather than just quantitative skills), to diagnose the successes and failures of any model.

## 16.3 MODEL BENCHMARKS

When a model or class of models is used repeatedly for ecological forecasting, the need quickly arises to define standards that can be used to assess model skill (Blyth et al. 2011; Luo et al. 2012). What separates these *benchmarks* from the quantitative metrics discussed in section 16.2 is that benchmarks are a set of standards that a community agrees upon for model assessment. Without agreed-on benchmarks, different groups would assess different variables at different sites using different metrics, making it impossible to compare how they're doing. Some benchmarks may involve checks that ensure that models adhere to basic physical constraints, such as conservation of mass or energy. Other benchmarks may involve testing models against an agreed set of data, which can be either observational or experimental in nature.

Benchmarks should concentrate on the focal variables that a community agrees are most important for a model to capture, or variables that highlight whether a model has captured a critical *process* correctly. The latter can be particularly valuable because many models can be calibrated to produce the correct outputs for the wrong reasons. In all cases, there need to be agreed upon skill metrics that allow different models to be compared quantitatively, as well as allowing the performance of individual models to be tracked over time. These may be basic statistics, such as RMSE or correlation, or they may be assessments of a model's ability to reproduce observed interrelationships among variables. When choosing among alternative skill scores for probabilistic forecasts, it is further desirable to include metrics that are proper and local (Bröcker and Smith 2007).

*Proper* refers to metrics that are maximized when the predicted probability distribution matches the true probability distribution, thus encouraging truthful forecasts. Typically, this means scoring probabilistic models using the same cost function as is used for inference and calibration (that is, the likelihood). *Local* refers to a skill metric that is calculated based on the values of the probability distribution around the actual outcomes/observations.

In addition to tracking improvement, a well-constructed set of benchmarks allows users to understand under what conditions a model is likely to perform well and allow model developers to identify model errors. In addition, benchmarks can allow a community to establish minimum standards for model performance for specific applications, which can be helpful in applied research to make ecological forecasts more defensible.

Because ecological forecasting is in its infancy for most ecological subdisciplines, there are very few examples of benchmarks for ecological models. Benchmarks have been proposed for land models (Blyth et al. 2011; Luo et al. 2012), but at the moment many of the proposed benchmarks are derived data products that involve a substantial number of assumptions to generate and often have poorly quantified uncertainties (for example, remotely sensed data products, spatially interpolated soil carbon pools). By contrast, attempts to benchmark models more directly against fine-scale observations can lead to a scale mismatch between models and observations.

Another challenge is that many models produce high-dimensional outputs (for example, maps, time-series) for many variables, which leads to a complex table of benchmark statistics. Such a table can itself be challenging to interpret, while attempts to generate composite scores across metrics can be somewhat ad hoc. In the atmospheric sciences, where the process of benchmarking is more established, skill is generally assessed with a smaller number of benchmarks that may not conform directly to the multitude of outputs of interest, but that experience has demonstrated are a good proxy for specific *processes* that are important for models to capture. It is also important that ecological benchmarks provide assessments that are nontrivial. For example, almost any model that is forced with meteorological observations that have a diurnal and annual cycle will produce diurnal and annual cycles in their outputs, as well as marked differences in seasonality between the tropics and the poles, so benchmarks that give high scores for reproducing such basic patterns may establish a minimum standard of performance but do little to assess whether models are working correctly. In this case the choice of an appropriate null model (section 16.5) can be critical for interpreting model skill scores.

## 16.4 DATA MINING THE RESIDUALS

Data mining approaches have seen increased usage as data sets have grown in size and dimension (Hoffman et al. 2011). In general, I am hesitant to embrace data mining as a useful method for forecasting due to the lack of process understanding and scientific hypothesis. In addition, most data-mining approaches depart from the probability-based approach to inference that is the focus of this book, and tend not to allow the same richness in the partitioning of sources of uncertainty and variability. That said, I do see a role for data mining as part of the model diagnostic process. In particular, as models and data become larger, visualizing (section 16.1) and interpreting (section 16.2) model-data residuals becomes more challenging, and it becomes harder to identify consistent patterns in model error. In contrast to mining the data itself, *mining the model-data residuals* may highlight conditions where our current process understanding is inadequate. Used in conjunction with process-based models, data mining may also provide a means for correcting model biases or adjusting model forecast uncertainties.

There are a wide variety of data mining approaches in use, with new approaches being rapidly developed, so it is impossible to give a complete overview of the techniques in use today. However, most of these approaches fall into a smaller set of categories, the most relevant to our model diagnosis problem being anomaly detection, clustering, classification, frequency-domain analyses, dimension reduction, and nonparameteric curve fitting. The value of anomaly detection in model diagnosis is fairly straightforward, as we are frequently much more interested in identifying large model errors than the background residual noise. Clustering and classification can be used to identify conditions when model error is notably higher or lower than average (figure 16.5, bottom right). For the subset of data-mining techniques that includes covariates, model inputs and settings can form part of the covariate space, since we are frequently interested in identifying not just when errors where high but also the conditions associated with differences in model error. Similarly, frequency-domain analyses, such as Fourier spectra and wavelet transforms, can identify the specific timescales associated with model error (Dietze et al. 2011; Stoy et al. 2013). Dimension reduction techniques, such as ordination, can be helpful when either the input or outputs of a model are high dimensional and correlated. Finally, nonparametric curve fitting techniques, such as general additive models (GAMs), boosted regression trees, and random forests, can assess model error as a continuous, smooth function of model inputs, rather than looking at error in discrete clusters.

## 16.5 COMPARING MODEL PERFORMANCE TO SIMPLE STATISTICS

A recurring theme in model assessment is how to construct an appropriate null model. This is relevant for interpreting model diagnostics, for establishing baselines for benchmarking, and for interpreting spectra and data mining model residuals. In assessing the performance of any process-based model, a critical test is to compare the model to what a simple statistical model would have predicted under the same conditions. In basic statistics we are accustomed to the idea of a null model, but frequently assume trivial null models, such as a constant mean or no relationship between two variables. The more appropriate question to ask in constructing a null

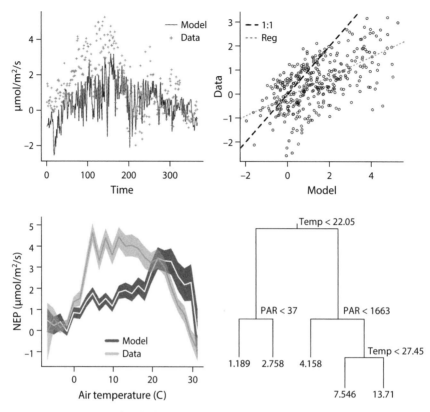

FIGURE 16.5. Diagnostics for daily mean net ecosystem productivity (NEP) from a simple ecosystem model compared to eddy-covariance observations. Top left: Time-series plot of model and data. Top right: Predicted/observed plot with 1:1 line and regression line, the latter illustrating model miscalibration. Bottom left: Functional response plot comparing observed and modeled NEP response to air temperature, illustrating a large difference in optimal temperature between model and data. Bottom right: Regression tree plot classifying how light (PAR) and air temperature (temp) affect the squared error between the model and data (values at leaves of the tree). Starting from the top, take the left branch if the listed inequality is true and the right branch if it is false. This shows that the highest errors are associated with the combination of high temperature and high light.

model is, "What would my forecast look like if I had no understanding of the ecological process?" While a constant mean may be sensible occasionally, more often than not our prediction would be that tomorrow is going to look a lot like today. As a quantitative forecast, this would be equivalent to the random walk state-space model that we discussed in chapter 8, which assumes that each time step is distributed around the previous time step, with some process error. However, in many cases a random walk may also be too trivial a null model because over longer timescales its uncertainty grows without bounds.

Imagine if we were predicting the weather in the era before weather models—we might assume that tomorrow might be a lot like today but that two months from now will be a lot like the average climate for that time of year. Indeed, even for my forecast for tomorrow, I might assume that the weather is going to be like today, but that it will also, on average, regress somewhat toward the climatic mean. Given the strong diurnal and annual cycle in weather data, we may further elaborate to include information across multiple time scales, such as assuming that an observation will be like the last observation, but also like an observation made at the same time of day on the previous day, and like the mean observation at the same time and same day in previous years, where the weighting of these factors may be found by empirically calibrating such a model to past observations. Overall, the take home is that in weather forecasting, the climatology provides an appropriate null model that process-based models can be evaluated against.

In the assessment of ecological forecasts, such comparisons to "ecological climatology" are rare, but represent an important benchmark. There's the additional challenge that most ecological systems being forecast are responding to directional environmental change and thus are nonstationary (Milly et al. 2008). This means that the climatological null model may itself no longer be an appropriate null. Indeed, this is an area that would benefit from additional research. Furthermore, if the history of weather forecasting is to be a guide, it is quite likely that many current models will fail to beat an appropriate null model, and may continue to do so for some time. This does not automatically demonstrate the futility of such forecasts, but rather illustrates that there is a learning curve in forecasting. However, the only way to progress in that learning curve is to get on the curve—we will not get better at making ecological forecasts until we start making forecasts.

## 16.6 KEY CONCEPTS

1. Model assessment should be approached from a hypothesis-testing perspective and look for creative visual and statistical assessments of model assumptions and processes rather than the tallying of scores.
2. A key goal of model assessment is to identify model structural errors and to distinguish these errors from other uncertainties (parameters, inputs, and so on).
3. Models should be tested against independent validation data, for output variables other than those used for calibration, and for their ability to capture functional relationships and emergent phenomena.
4. Benchmarks define standards that can be used to repeatedly assess model skill.
5. Data mining model-data residuals may highlight conditions where our current process understanding is inadequate.
6. Model diagnosis benefits from the construction of an appropriate, nontrivial null model, such as simple statistical models that lack process understanding.

## 16.7 HANDS-ON ACTIVITIES

https://github.com/EcoForecast/EF_Activities/blob/master/Exercise_11_Model Assessment.Rmd

- Model sanity checks
- Error statistics

- Climatology
- Taylor diagram
- Timescales
- Data-mining residuals
- Capturing functional responses

# 17

## Projection and Decision Support

SYNOPSIS: *Forecasts into the future are frequently done conditional on a specific scenario about how factors outside the purview of the model will evolve, such as climate, land use, and policy. Scenarios also allow us to explore the possible impacts of different actions. Outputs of one forecast are frequently inputs into another, such as policy analyses and decision support models. This chapter discusses techniques for developing scenarios and providing decision support.*

BASED ON THE material covered thus far in this book, hopefully you now have a general idea of how ecological forecasting can work. For a given question facing the forecaster, models need to be developed to forecast the quantity of interest. Data need to be collected and collated (chapter 3) to calibrate the model (chapter 5) while dealing with a wide range of uncertainties (chapters 2 and 6). The quantity of interest itself is frequently treated as a latent variable (chapter 8), and often multiple sources of data are available to directly or indirectly inform its estimation (chapter 9). Models can be analyzed to determine their sensitivities and uncertainties, which feeds back on observational design decisions of what to measure or monitor (chapter 11). Forecasts can be made that propagate multiple sources of uncertainty (chapter 11) and monitoring data feeds back to refine these forecasts through time in what is known as the forecast cycle (chapters 13 and 14). Finally, models need to be verified to assess their performance and identify structural errors (chapter 16). For good science and policy, it is essential that all these steps be open and repeatable, and in some cases workflows may be developed to automate the forecast cycle into "operational" forecast systems (chapter 4).

So what is left to cover? Let's return to our original definition of ecological forecasting in chapter 1:

> the process of predicting the state of ecosystems, ecosystem services, and natural capital, with fully specified uncertainties, and is contingent on explicit scenarios for climate, land use, human population, technologies, and economic activity. (Clark et al. 2001)

From this it should be clear that I've spent the bulk of this book focused on the process of predicting with fully specified uncertainties, but have said far less about scenarios. So what are scenarios and when do we need them? To answer this question we need to step back and ask why we are making a forecast. Sometimes forecasts are done to answer basic science questions, but in many cases forecasts support management and policy decisions. So how can ecological forecasts enter into decision support? These questions are the subjects of this chapter.

## ▉▉▉ 17.1 PROJECTIONS, PREDICTIONS, AND FORECASTING

When thinking about the forecasting problem more generally, MacCracken makes a useful distinction between predictions versus projections (MacCracken 2001). A *prediction* is a "probabilistic statement that something will happen in the future *based on what is known today*." The most familiar example of prediction is weather forecasting. By contrast, a *projection* is a "probabilistic statement that it is possible that something will happen in the future" *conditioned on boundary condition scenarios*. Intergovernmental Panel on Climate Change (IPCC) climate projections are the analog to weather predictions—they are long-term forecasts conditioned on specific scenarios of emissions, economic activity, population, and land use. These boundary conditions (also known as drivers) are not themselves based on predications from each of these sectors, but representative "what if" statements.

To give a more ecological example, imagine we are forecasting the impact of land use change on some aspect of an ecological system (for example, invasive species, endangered species, carbon flux, river nutrient load). Any ecological forecast would thus be made conditional on land use change as a boundary condition that changes over time. If a colleague has a land use change model that has been used to *predict* land use change with uncertainties, then this driver and its uncertainties can be incorporated into your forecast using the techniques discussed in chapter 11 (for example, Monte Carlo simulation, ensemble forecasts, and so on). In this case you are trying to integrate over the land change uncertainty and your forecast would be a *prediction*. Next, consider the case where a different colleague has generated different *scenarios* of land use change—*different stories about how the future might play out*. In this case you are making a distinct probabilistic *projection* for each scenario. However, the different scenarios are not meant to be random samples from the set of plausible futures. Therefore, unlike an ensemble analysis, you would not want to average over different scenarios, but rather evaluate them individually.

In the literature the term *scenario* can sometimes be used to mean different things, so for any particular context it is important to be clear about what one is referring to (Peterson et al. 2003). In some cases scenarios refer to a set of *storylines* describing different plausible futures (Biggs et al. 2010). These plausible futures are not part of an exhaustive set or formal probabilistic analysis and thus cannot be assigned weights or probabilities. Projections in this case are often meant to be broadly informative rather than tied to a specific decision making process. The IPCC climate change scenarios would fall into this category (Moss et al. 2010). Next, *scenario* might be used as a synonym for *alternative* in decision support. When making ecological forecasts to support a specific management or policy decision, there are usually alternative actions being considered, and the role of the ecological forecaster is thus to make projections under each alternative (see section 17.2).

Third, scenarios can be used to address rare or *low-probability events* that could have large impacts on a system. Such events often fall outside the scope of models and thus are not captured by routine predictions. In other cases, it's not possible to assign them reliable probabilities, and thus they similarly can't be included in predictions. As discussed in chapter 2, these are the "failures of imagination" that very often lead to forecast failures—what Rumsfeld called the "unknown unknowns" or what Taleb calls "black swans" (Taleb 2007). Projections of rare events are often used to "war game" strategies for how to cope with them or to develop alternatives that are either robust to such event or are "bet hedging" (for example, devoting a small fraction of resources to activities that are sub-optimal or even counterproductive under most conditions but protect against the worst case). They can also be used for purely scientific purposes (for example, gaming out implications for competing hypotheses). Similar "worst case" scenarios may also enter into decision support when comparing different alternatives.

## 17.2 DECISION SUPPORT

Fundamentally, decisions are always made on the basis of what we think the future will look like, not on what has occurred in the past. Therefore decision support is at its heart a forecasting problem, not a monitoring problem, though as we have seen throughout this book, monitoring data are essential to developing good forecasts. In addition, decisions are always made with uncertainty—as we will see later, an accurate representation of the relevant uncertainties is critical to good decision making. Underestimates of uncertainty can lead to overconfident decisions, while overestimates can lead to excess caution.

There are numerous different approaches by which decisions about ecosystems and ecosystem services are actually made in the real world. Even when approached formally, there are many competing theories on decision making that are beyond the scope of this book to describe. The goal of this chapter is not to turn the reader into a policy analyst, but to increase the average ecologist's background knowledge about decision support. If we better understand what we need to provide decision makers, we are in a better place to think about how ecological forecasts can enter into the decision making process. This chapter is intended to give you a starting point to work with experienced policy analysts, political scientists, and economists.

In this section I will focus on one particular approach to decision support based on a consequence table as this illustrates many key ideas about how forecasts support decisions (Gregory et al. 2012). A *consequence table* summarizes how different alternatives perform for different performance measures being used to evaluate the alternatives (table 17.1). In any particular decision context the exact steps and approaches may be very different, but my hope is that the underlying concepts will be useful to ecologists providing input on environmental decisions.

Decision support (figure 17.1) begins with determining the objectives of the decision. The rows in the consequence table represent the different *objectives* being considered—the things that the decision maker, stakeholders, or larger community have decided that they want to either maximize (for example, biodiversity, productivity, economic returns) or minimize (for example, risk of extinction, invasion, or infection; loss of jobs, property value, or cultural heritage). Next, for each objective, *performance measures* are selected that either describe or are proxies for these objectives. After that different *alternatives* are generated—these are a discrete set of mutually exclusive choices available to decision makers. The columns in the consequence

TABLE 17.1 Example Consequence Table for Alternative Power Supply Options
to a Remote Community

| Attribute | Units | Alt 1 | Alt 2 | Alt 3 | Alt 4 | Alt 5 |
|---|---|---|---|---|---|---|
| Unit energy cost | $/MWh | 149 | 114 | 110 | 124 | 108 |
| GHG emissions | Kilotons/yr $CO_2$e | 31 | 8 | 8 | 16 | 8 |
| Local air emissions | Tons/yr (PM10) | 16 | 17 | 21 | 9 | 24 |
| Land area | m² (×1000) | 29.7 | 16.8 | 4.6 | 19.6 | 3.1 |
| Aquatic area | m² (×1000) | 8 | 24 | 0 | 35 | 20 |
| Construction jobs | Person-years | 75 | 119 | 105 | 96 | 119 |
| Permanent jobs | FT equivalent | 49 | 81 | 83 | 76 | 84 |
| Noise | (0 = best, 10 = worst) | 6.7 | 3.1 | 3.7 | 3.6 | 3.9 |
| Visual impacts | (0 = best, 10 = worst) | 1.5 | 2.2 | 2.8 | 1.4 | 2.2 |
| Food harvesting area | (0 = best, 10 = worst) | 1.5 | 0.9 | 0.5 | 1.4 | 0.2 |
| Sustainability/innovation | (0 = best, 10 = worst) | 0 | 0.3 | 0.5 | 0.7 | 0.3 |
| Sustainability/innovation | % peak renewables | 12% | 22% | 23% | 12% | 25% |

Source: Reproduced from www.structureddecisionmaking.org.

table represent different alternatives under consideration. Once the dimensions of the decision process are laid out, the next task is to forecast the *consequences* of the alternative decision choices. The cells of the table contain the individual consequences for a given alternative and objective. For ecological performance measures these would be the outputs of our ecological forecasts. For other objectives consequences may come from economic projections, surveys, or even qualitative descriptions. The different performance measures will thus typically have different natural units. Unlike a cost-benefit analysis consequences are not converted to a common unit (for example, monetizing different ecological or cultural values based on willingness to pay). The filling in of consequences focuses specifically on *information*, not the values we attach to different objectives, and thus ranks and preferences are not used as performance measures.

While the construction of the consequence table is information focused, it is important to remember that decision making is ultimately not a technical question, but a question of *values*: beliefs, priorities, preferences, tolerance for risk, and time discount (value of short- versus long-term benefits). Factual information plays a critical role, but hard decisions ultimately come down to value-based choices about how decision makers resolve the trade-offs among different objectives. The ultimate goal of the consequence table is to highlight and evaluate such value-based *trade-offs* and make them explicit—to focus decision making on preferences among alternatives rather than either marginalizing values or letting them dominate the discussion. It aims to address values constructively.

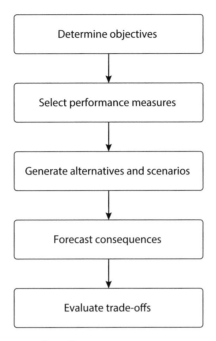

FIGURE 17.1. Decision support flowchart.

The remainder of this section will focus in greater detail on the steps involved in constructing a consequence table (figure 17.1), and ways we can analyze this information to assist in decision making. Much of this material is derived from Gregory et al.'s book *Structured Decision Making*, which provides an excellent reference on decision support (Gregory et al. 2012).

### 17.2.1 Objectives and Performance Measures

The first step in constructing a consequence table is to *determine the objectives*. Objectives consist of a few words that summarize something that matters to the stakeholders and the desired direction of change (for example, maximize net carbon storage). Different stakeholders need not agree on how much they value a specific objective, but inclusion does validate that a specific objective *has value*.

Objectives are context-specific—they are not statements about universal values, but rather their value in judging among alternatives in this one specific question. Objectives should indicate the desired direction of change (for example, increase employment), not prescribe a specific target (for example, create 500 new jobs) as the latter is arbitrary and its binomial nature (success or failure) obfuscates trade-offs. Where decisions involve legal thresholds (for example, contaminant concentrations), objectives should be expressed in terms of *probability of meeting a standard*. Objectives should also be focused on ends (for example, protect biodiversity) not the means (restore habitat) as the generation of alternatives is aimed at exploring the means. Objectives should not be assigned weights at this stage, which is focused on understanding what matters. Weights are assigned only at the end, once consequences have been estimated, inferior alternatives eliminated, insensitive measures dropped, and key trade-offs identified.

Once objectives are articulated, *performance measures need to be selected*. In general, performance measures should quantify objectives in units that are appropriate for that objective and should not be monetized unless it is normal to think about the objective in monetary terms. Performance measures fall into three categories: natural measures, proxies, and constructed measures. *Natural measures* directly describe outcomes and are clearly preferred when they are available. *Proxy measures* are indirect indicators of things that are difficult to measure. Proxies are abundant in ecology (for example, composition as a proxy for ecosystem function or health, remote sensing as a proxy for productivity, and so on), but their use often hides uncertainties. In addition, there are always concerns that the correlations used to build proxies won't hold over time, in new locations, or under novel conditions. Finally, *constructed measures* are useful for objectives that are hard to measure or lack any natural measurement. Everyday examples would be the 0-to-5 star scale frequently used in online shopping, restaurants, and hotels. While subjective, people are actually quite accustomed to assigning constructed quantitative measures to qualitative phenomena. Because simple rating scales lack any sort of calibration (different individuals may not mean the same thing with a 3-star ranking) a preferred alternative is to use *defined impact scales*, which combine a numerical score with a description of each ranking that articulates the key factors. Another challenge with constructed scales is their nonlinear nature. An alternative that scores an 8 out of 10 on measure isn't necessarily 4× better than a 2. If constructed measures need to be used quantitatively, then it becomes necessary to rigorously elicit the relationship between rank and value (section 17.2.4).

Regardless of whether they are natural, proxy, or constructed, uncertainties exist for all measures and these need to be carefully documented and reported. For most objectives the absolute performance of a metric is not what's valued, but the relative performance compared to other alternatives. Because a lot of the assumptions and uncertainties in a forecast will be shared among all alternatives, relative measures often have lower uncertainty than absolute measures because some uncertainties will "cancel out." This point reinforces that all measures have to be treated consistently; *all values for a single performance measure (row) need to be calculated the same way with the same assumptions*. It also reminds us that all performance measures are forecast projections, not direct measures or monitoring.

Finally, it is important to refine objectives and develop performance measures iteratively. Those involved in decision support should work with decision makers to verify that measures are useful and understandable. Estimation of consequences can start with rough calculations and expert opinion, in lieu of expensive and time consuming modeling, to assess what is worth measuring and modeling (chapter 11) and eliminate measures that will be insensitive (section 17.2.5).

### 17.2.2 Alternatives and Scenario Development

Generating scenarios is central not just to decision making but to all projections. The following discussion focuses on the development of alternatives in the context of decision support, but recall that decision alternatives are just one class of scenarios. That said, generating scenarios in other contexts is very similar, and I will try to note any nontrivial differences that would occur when developing general purpose scenarios or war-gaming.

One of the challenges in developing alternatives is the creative exploration of the full space of possible choices. This occurs due to a suite of cognitive biases—people are not rational and frequently employ simplifying rules of thumb that can sometimes become traps (Kahneman 2013). One of the most common is *anchoring*, whereby both the alternatives considered and the value assigned to alternatives are expressed relative to an initial impression, frequently the status quo. Because of this, a common strategy is to start with *bookend* alternatives. Bookends are constructed by considering the best and worst cases for each objective independently. Bookends shift the focus to the range, rather than a preanalysis preferred outcome, and help validate the ideals and fears of different stakeholder groups. Other biases to watch for are the *representativeness* bias (stereotyping), *availability* bias (giving more weight to recent examples), and the *sunk cost* bias (justifying and protecting earlier choices—as is often said, if you find yourself in a hole, stop digging).

At an early stage, the focus should be on generating and exploring alternatives, not trying to reach a consensus, as early consensus limits exploration of alternatives, controversies, and trade-offs. Thus, another bias to watch for is *groupthink*, the premature convergence on one alternative without sufficient analysis, which can occur because of both a natural human tendency to not want to spoil an emerging agreement and the influence of strong or vocal personalities. Thus, once bookends are established, it is important to brainstorm a wide range of alternatives, look for creative solutions, and make sure that minority views are represented. It is not uncommon for alternative development to go through multiple rounds that hybridize different alternatives (that is, borrow the best parts) to search for win-win alternatives in place of apparent trade-offs. Finally, when generating alternatives it is important to realize that *any decision is only as good as the set of alternatives considered*. All judgments and evaluations are relative to that set. People are unlikely to be satisfied with a tough decision if there are alternatives that were not considered.

So how many alternatives should be considered? The first constraint on this is one of resources. If you start evaluating multiple levels of multiple factors this may quickly lead to hundreds to thousands of alternatives. Time, money, manpower, and computation may impose real, practical limits on this type of search, which is in many ways analogous to a global sensitivity analysis (chapter 11). Even if such a large number of alternatives could be computed, ultimately these would need to be narrowed down (using the approaches in section 17.2.5) to 4 to 12 alternatives that a group of stakeholders would be able to meaningfully discuss and evaluate, and then only 3 to 4 that would be presented to decision makers or end users (Schwartz 2005). Having too many alternatives/scenarios under consideration leads to a "paradox of choice"—keeping track of the alternatives and the differences among them becomes difficult. When possible, it is also preferable to choose an even number of alternatives/scenarios, as an odd number can lead to anchoring (for example, assuming a middle scenario is the mean). It is at this "winnowing" stage that alternatives should be assigned descriptive names. Names need to be neutral so that alternatives are judged by their performance measures, not their names (that is, avoid names such as "eco-friendly" or "pro-business").

Overall, alternatives need to be complete, comparable, and internally consistent solutions. This means that all alternatives address the same problem, evaluated over the same time, to the same level of detail, and with the same assumptions and performance metrics. This also implies that they need to be mutually exclusive (that is, you can't choose more than one of them) otherwise they are not complete. Finally, alternatives have to include enough detail that they can be used to drive projections. Therefore the level of detail required by a model can influence the detail required in specifying the alternatives.

An excellent example of scenarios that most ecologists are familiar with are the four Representative Concentration Pathways (RCPs) the IPCC relies on for its climate change scenarios (Moss et al. 2010). These are a manageable and even number of scenarios with meaningful but neutral names (RCP8.5, 6.0, 4.5, and 2.6 refer to their W m$^{-2}$ forcings). They explore a wide range of possible outcomes (though not bookends), and are detailed enough to drive climate models.

### 17.2.3 Consequences and Uncertainties

Given the set of objectives and performance measures as your rows and the set of alternatives as your columns, at this point the structure of the consequence table is laid out and the next task is to estimate the values of each cell—the *consequences*— and their uncertainties. This is when studies are conducted to generate estimates for the different performance measures. Some of these estimates will come from expert elicitation (see box 5.2) or data and literature synthesis (chapter 9), especially if there is an initial round of analysis aimed at reducing the scope of the problem or identifying the dominant uncertainties. *This stage is also where the bulk of ecological forecasting will occur in a decision support context.*

When estimating consequences, the inclusion of uncertainties is critically important. At a high level we can divide uncertainties into those that are *linguistic* versus *epistemic* (Regan et al. 2002). *Linguistic uncertainties* are those associated with language and communication, such as vagueness, ambiguity, context dependence, underspecificity (unwanted generality), and indeterminacy (change in meaning over time). Reducing linguistic uncertainty is a first priority. Because of this consequences should focus on facts, not values, and shouldn't be reported as pros and cons (or other linguistically ambiguous or value-laden rankings—for example, high/low).

*Epistemic uncertainties* are those associated with our knowledge, and have been the focus of much of this book. As we've discussed throughout this book, these uncertainties should be estimated as probabilities. Partitioning these uncertainties (chapter 6) can also be extremely helpful, especially if it is possible to separate uncertainties into those that are *reducible* via additional data collection (for example, parameter uncertainty) and those that are *irreducible* (for example, natural variability, stochasticity). It is also useful to identify uncertainties associated with our ignorance, by which I mean where we don't understand the nature of an uncertainty or the cause and effect mechanisms involved in a process. While consequences should generally be reported probabilistically, there are a few cases where this isn't always required. For example, when uncertainties are overwhelmingly large it is often more productive to evaluate how different alternatives perform under a discrete (and unweighted) set of scenarios. The most familiar example of this is again the IPCC Representative Concentration Pathways, which are discrete scenarios used in lieu of attempting to fully integrate all the uncertainties in technology, economics, population, and land use over the next hundred years. That said, ecological forecasts of climate change impacts and feedbacks can and should be done probabilistically *conditional on* each RCP scenario.

As noted earlier, consequence estimation is frequently done iteratively, in which case the uncertainty analysis techniques from chapter 11 can be used to identify the most important sources of uncertainty. While all consequences will have uncertainties, in practice only a subset of these will matter. Further data collection, refinements of models, and improvements of models should *focus specifically on those terms that affect the outcome of the decision.*

When generating estimates with uncertainty, careful thought should be given to *how uncertainties are reported.* In particular, you probably don't want to fill the consequence table with long lists of summary statistics, and many end users may not want (or need) the full PDFs. On the flip side, you also don't want to just report the expected value (that is, mean) or most likely outcome (mode). Common scientific approaches to uncertainty reporting, such as standard errors or confidence intervals, may also not be the best representation of uncertainty for decision makers. For example, most end users who have not been trained in statistics (and many who have) will interpret all values within a confidence interval as equally likely. Even simple things, such as how probabilities are reported, can lead to different levels of understanding and perceived risk. For example, a relative frequency (for example, 1 in 20) is easier for most people to use, but is perceived as a higher risk than the same value reported as a percent (for example, 5%). Similarly, for extremely low-probability events the severity of the consequence often becomes the focus and the probability is frequently ignored because it is hard for us to visualize differences among low probabilities (for example, 1 in 1 million versus 1 in 1 billion). When interpreting consequences, it can be helpful to provide context for the values reported. For example, how do the values reported compare to other similar systems or some baseline? Any baseline needs to be chosen carefully because people do not perceive gains and losses as having the same *value* even if they have the same magnitude (for example, a 10% loss is perceived, on average, as twice as bad as a 10% increase) (Berger 1985; Kahneman 2013).

This idea of how gains and losses are perceived is tightly coupled to the idea of *risk tolerance.* Risk tolerance will be discussed in greater detail in the next section, but understanding how risk is perceived plays a large role in how uncertainties should be reported. For example, because risk perception is asymmetric (losses are worse than gains), one useful approach is *downside reporting,* which involves reporting the value of a specific performance metric under a "worst plausible case." The worst plausible case is not necessarily the absolute worst case, which may have a vanishing small probability, but rather a specific predefined lower probability, such a 1 in 10, 1 in 20, or 1 in 100, that would be rare but not unexpected. Another similar approach to reporting risk would be the probability of exceeding some regulatory, or agreed upon, threshold. Downside reporting and exceedance probabilities are useful for managers because the consequences of being wrong are often much more detrimental than the consequences of underestimating a benefit.

The consequences of being wrong also vary considerably with the nature of the decision itself. Management decisions that are made frequently and under similar conditions may be more "risk neutral," because the few cases that perform worse than average would be expected to be balanced by times when the performance is better than average. By contrast, for decisions that are one time and/or high risk, decision making will be more risk adverse. In this case downside reporting can assist with developing alternatives that are precautionary, robust, or adaptive against uncertainties.

*Precautionary alternatives* are ones that apply the precautionary principle embodied in the United Nations (1993) Rio Declaration on Environment and Development: "Lack of full scientific certainty should not be used as a reason for postponing cost-effective measures to prevent environmental degradations." However, it is worth noting that being precautionary on one objective often involves trade-offs that cause other objectives to become wasteful or inefficient. In decision support it may be more useful to develop alternatives that explore different degrees of precaution. *Robust alternatives* are ones that perform adequately over a wide range of uncertainties and plausible futures. The cost of robust alternatives is that they typically don't perform as well under any specific set of conditions. By contrast to robust alternatives, brittle alternatives are ones that are optimal but sensitive to deviations. Finally, *adaptive alternatives* are those that learn as they go, refining approaches as additional data is collected and uncertainties reduced over time. Adaptive management, in particular, should involve active experimentation to resolve among competing hypothesis (Walters 1986) and the use of iterative forecasting approaches to update analyses as new data becomes available (chapters 13 and 14). Adaptive alternatives have many advantages, but do incur costs in both monitoring and time.

### 17.2.4 Utility: Value and Risk Tolerance

At this stage in decision support the full consequence table has been populated with forecasts and the primary remaining task is to evaluate the results so as to identify trade-offs and eliminate inferior alternatives. However, before we dive into trade-offs (section 17.2.5), I want to make a brief detour to discuss how human values and risk tolerance can be quantified and how we can use this to formally account for the impact of forecast uncertainty in decision support.

Both values and risk tolerance are central to how decisions are made, from simple everyday choices to international negotiations. For example, when choosing what to have for lunch there's a spectrum of preferences among individuals, low-level risk about whether a particular meal will be enjoyed, and lower probability but higher consequence risk about whether a meal will make us sick. There's also a range of risk preferences among individuals, leading some individuals to seek out variability and others to settle into routine. Economists represent these concepts of value, preference, and risk tolerance as *utility*. While the previous sections focused on separating knowledge and values when constructing a consequence table, utility is the focus when evaluating alternatives.

Things that are rare (for example, individuals in an endangered population) are generally valued much more than things that are common. As the amount of something we have increases the "per unit" value (what economists call *marginal utility*) tends to decrease, even if the cumulative value increases. Mathematically this means that utility is a nonlinear, and typically concave, function. Therefore, whenever we make decisions with uncertainty (which is all decisions), Jensen's inequality rears its ugly head once again. The implication of this is that the expected value for any consequence (also known as the mean) is, by itself, a poor metric for decision making, even if it does describe the most likely outcome. More formally (equation 11.2), it means average utility will always be less than the utility of the mean, and *utility declines directly as a function of forecast uncertainty*.

In addition to value being nonlinear, there is considerable variability in the population with regard to risk tolerance. Some individuals will be risk-neutral, but most will be risk adverse (tend to value protecting against losses more than gains), and a few will be risk tolerant (willing to take large gambles for large rewards). In most cases risk tolerance does not map neatly onto any stakeholder group or political affiliation. Given that human decisions can be highly sensitivity to risk, the accurate quantification of uncertainty (neither under- nor over-estimating) is absolutely essential to ecological forecasts that are tied to decision making. That said, decisions definitely don't have to be free of uncertainties—numerous individuals involved in decision support have reiterated to me that *"better than 50:50" is the only threshold for ecological forecasts to be useful for decision making.*

### 17.2.5 Utility Models

While there are many cases where utility can be addressed through performance measures (for example, downside reporting), there are times when it can be advantageous to address utility more formally. Utility models aim to quantify values, preferences, and risk tolerance. For example, consider the case of a one-time bet where an individual is given the choice between a guaranteed reward (for example, $500) versus a gamble (for example, a 50% chance of winning $1000). Someone who is risk neutral would view these two choices as equivalent since they have the same expected value. By contrast, someone who is risk adverse would prefer the guaranteed reward, while someone who is risk tolerant would prefer the gamble. But *how much* more are these alternatives preferred? Through expert elicitation (chapter 5), an individual could be quizzed about alternative bets to find their personal views of what constitutes a "fair trade" (economists refer to this as *indifference*). For example, a risk-adverse individual may be willing to accept $300 in hand as a fair trade for a 50% chance of winning $1000, even though the expected outcome of the latter is higher (Berger 1985). More formally, this would imply that the utility, $U$, of $1000 is twice that of $300 (figure 17.2) because their expected utilities $E[U]$ are the same, $E[U] = Pr(y)*U(y) = 1*U(\$300) = 0.5*U(\$1000)$. Furthermore, the utility of losses are not the same as the utility of gains—this same hypothetical individual may value a 50% chance of losing $1000 different from a $300 guaranteed loss. By eliciting a set of such fair trades a *Utility function, $U(y)$,* can be approximated. For example, the utility of money is often approximately logarithmic. The units of Utility do not have to have some absolute meaning (that is, units can be relative), because they are used to examine *relative* preferences. That said, a very common metric of utility in environmental policy analysis in general, and cost/benefit analysis in particular, is to express utility in terms of monetary *willingness to pay* for some environmental good or service.

A primary use of utility functions is to compare the expected Utility for different performance measures. If $P(y)$ describes the probability distribution of a specific performance measure for a specific alternative under consideration, then the expected utility of action $a$ is given as

$$E[U] = \int U(a,y)P(y)dy$$

Unlike the expected value of the performance measure, $E[y]$, which is inherently risk neutral, expected utility captures risk tolerance. Therefore, it can be used to assess how we *value* different alternatives in the face of uncertainty. For example, $E[U]$ may help clarify the relative merits of a high-risk/high-reward alternative and a more precautionary alternative. Furthermore, utility theory not only explains why we are

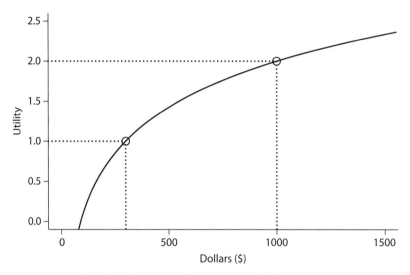

FIGURE 17.2. Example utility function. The meaning of utility is primarily a relative measure so the units can be arbitrary. In this example the reference value of $300 is used as a baseline ($U = 1$) for estimating the utility of $1000.

more risk neutral in the face of repeated decisions but also allows us to formally calculate how much our utility changes. For example, if I were given *two* independent opportunities to make the gamble between a guaranteed $300 return and a 50% chance at $1000 then the utility of the nongambling strategy is not $1*U(\$300) + 1*U(\$300) = 2$ but rather U($600). For the gambling strategy there are three possible outcomes: $P(\$0) = 0.25$, $P(\$1000) = 0.5$, $P(\$2000) = 0.25$. Therefore $E[U] = P(\$0)U(\$0) + P(\$1000)U(\$1000) + P(\$2000)U(\$2000) = 1 + 0.25[U(\$0) + (\$2000)]$. Overall, because $U$ is nonlinear there's no guarantee that U($600) = 1 + 0.25[U(\$0) + (\$2000)]$. More generally, under repeated independent decisions the variance tends to decline (that is, basic sampling theory), and therefore the $E[U(x)]$ moves closer to $U(E[x])$, which corresponds to risk neutrality. Thus, *as a decision becomes more routine, we become progressively more risk neutral.*

One thing to note about utility functions is that they reflect an *individual's* preference, so the preferences of a community would result in a community of utility functions and thus a distribution of expected utilities. Because of Jensen's inequality, one cannot simply construct an "average" utility function. Finally, utility functions themselves are empirically constructed, and thus subject to uncertainties, which can be valuable to explore through sensitivity analysis and/or propagated into the expected utility calculation.

### 17.2.6 Trade-offs

The previous section on utility illustrates how uncertainty within a *single* performance measure can create trade-offs between risk and performance. For example, a high-risk, high-reward alternative could have a higher mean performance than a conservative alternative, but a lower expected utility. This section will focus instead on decision making in the face of multiple competing objectives. In some circumstances, through creative alternative generation and a bit of luck, it is possible to find win-win alternatives. In this case decision support is straightforward. However, more commonly we are faced with trade-offs among different alternatives.

In the event that there is not a clear winner, decision support focuses on the "winnowing" of alternatives. The approach is to iteratively improve the set of alternatives by (1) deleting bad alternatives and insensitive performance measures; (2) refining our understanding of the key trade-offs; and (3) adding new alternatives that address these trade-offs. The goal of these analyses is to reduce the set of alternatives to those that really highlight the important trade-offs and risks. At that point decision support stops being a technical exercise focused on information, and becomes an actual decision for managers and policy makers, which is focused on values.

In the context of decision support *dominated alternatives* are those that lose on all measures when compared to another alternative. These are the alternatives that are universally "bad" and should be eliminated. When working with a small number of alternatives, winnowing can be done by hand. This exercise can be valuable so that all stakeholders are on board with dropping alternatives, and because there may be opportunities to salvage the best parts of rejected alternatives or address their deficiencies once the source of their failings becomes clear. Elimination by hand is done based on pairwise comparisons. Some comparisons will be *strictly dominated*, which means that the inferior alternative performs worse for every measure. Other comparisons will be "*practically*" *dominated*, which means that there are actually trade-offs but they are judged to be negligible. The uncertainties in different measures can also play a role in determining "practical" dominance, but it is essential to remember that *this is* not *a matter of checking for overlapping CI*. In decision making, uncertainty translates into risk. If option A had a higher mean but wider CI than option B, this represents a genuine management trade-off, not a "nonsignificant" difference. As dominated alternatives are removed, *performance measures (consequences) that are insensitive to alternative choice can be dropped from consideration since they no longer affect the outcome.*

When dealing with hundreds and thousands of alternatives, or alternatives that form a continuum rather than discrete choices, elimination by hand is clearly infeasible. In this case one can rely on numerical algorithms for multi-objective optimization, also know as *Pareto optimization*. Rather than finding a single optimum, Pareto optimization seeks to clarify the trade-off front (also known as the *Pareto Front*), which consists of the set or manifold of Pareto optimal solutions (figure 17.3).

*In many cases decision support stops here, with a narrowed set of alternatives and a clear identification of the trade-offs among them.* The techniques to reach this point are strictly knowledge-based and do not require elicitation of the relative value of different objectives. Further refining the set of alternatives considered requires assessing the values of a community or group of stakeholders. Making a decision in the face of trade-offs among performance measures depends upon understanding how different individuals weight the importance of different objectives. As mentioned earlier in this chapter, such weights should not be assigned *a priori*, but only after performance measures have been calculated, dominated alternatives eliminated, and insensitive objectives dropped. This allows for greater focus on the practical performance of different alternatives, rather than philosophical generalities. It also avoids unnecessary debates over the value of objectives that may end up dropped or unrelated to the key trade-offs.

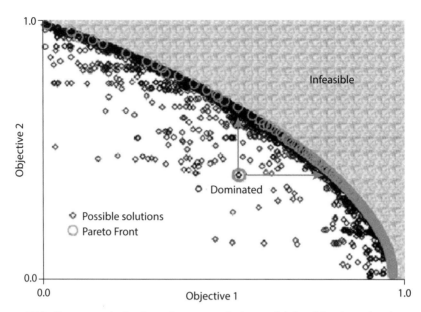

FIGURE 17.3. Pareto optimization aims to optimize multiple objectives simultaneously, eliminating alternatives that are dominated (perform worse on all measures), and thus identify the trade-off surface across objectives (the Pareto Front). Reproduced from Fingerhut et al. (2010).

In lieu of a full-blown utility analysis, when comparing alternatives it can be useful to employ multiple methods to assess whether decisions are robust. One simple first approach would be to have stakeholders identify which alternatives they endorse (enthusiastically support), which they are willing to accept, and which they actively oppose. This approach may allow the further elimination of alternatives that lack support, and may obviate the need for numerical rankings or weights. Similarly, analysis of the consequence table may identify *no-regret actions*—features that are common to all alternatives along the Pareto Front that can be implemented immediately without disagreement.

One method of assigning weights to different objectives is *swing weighting*. Swing weighting begins by imaging an alternative that performs *worst* on all metrics and then asks the question, "If you could improve one metric (switch it to the best score), which one would it be?" This objective is labeled 1 and given 100 points. The process is repeated to identify the second most important objective, and this is assigned points relative to best (that is, if the second objective is ½ as important it is given a score of 50). After all objectives have been scored, then the weight of each is calculated as

$$w_i = \frac{p_i}{\Sigma p_i}$$

where $w_i$ is the weight of the $i^{th}$ objective and $p_i$ is the points assigned to that objective. An alternative weighting approach would be to ask individuals to rank all objectives and then to assign points to each objective, with the top objective again starting with 100 points. Regardless of the weighting method used, it is important to have individuals construct weights, rather than trying to reach a consensus or calculate mean weights as a group, and then to look at the distribution of scores across individuals. Remember, the goal of decision support is not to reach a group consensus but to inform decision makers about how different trade-offs are viewed.

Finally, sensitivity analyses (chapter 11) should be performed to determine if choices are robust or if they are sensitive to small changes in weights. Similarly, *critical value* analyses assess *how much a performance measure would have to change to change the decision*, and from there to calculate the *probability of crossing this threshold*. When assessing the uncertainty in performance measures this is often more informative than a standard sensitivity analysis. More generally, these types of analyses may avoid the need to invest time and effort into the construction of formal utility models. Furthermore, assessing sensitivities may identify additional robust, no-regret actions. Alternatively, it may drive targeted data collection prior to decision making or refine adaptive management strategies. Given the costs of adaptive management, adaptive strategies make the most sense in cases where the refinement of uncertainties over time (chapters 13 and 14) would change outcomes or choices.

## 17.3 KEY CONCEPTS

1. Fundamentally, decisions are about what will happen in the future, in response to our choices, rather than about what's already happened in the past.
2. Predictions are forecasts that integrate over driver uncertainties, while projections are forecasts conditioned on external "what if" scenarios.
3. Scenarios can represent future storylines, management alternatives, or low-probability events.
4. Objectives summarize something that matters to the stakeholders and the desired direction of change.
5. Performance measures quantify objectives in units that are appropriate for that objective.
6. Decision support must address a wide suite of cognitive biases in how both alternatives and probabilities are generated and perceived. This can have large impacts on how alternatives are framed and uncertainties reported.
7. Any decision is only as good as the set of alternatives considered.
8. Precautionary, adaptive, and robust alternatives can help guard against risk and uncertainty.
9. Within decision support, the bulk of ecological forecasting will occur in estimating the consequences of different alternatives.
10. Utility functions are used to quantify the nonlinear relationship between quantities and values, and can give insight into risk tolerance.
11. Human decisions can be highly sensitive to risk—average utility declines directly as a function of forecast uncertainty. The accurate quantification of uncertainty is essential to guard against decisions that are overconfident or excessively precautionary.
12. The final stage of decision support is to eliminate bad alternatives and insensitive performance measures in order to identify key trade-offs.
13. The only "bad" trade-offs are the ones we make unknowingly, or without fully appreciating their implications.

## 17.4 HANDS-ON ACTIVITIES

https://github.com/EcoForecast/EF_Activities/blob/master/Exercise_12_Decision
Support.Rmd

- Stakeholder role-playing game
- Construction of a consequence table
- Fermi estimation
- Evaluating trade-offs

# 18

## Final Thoughts

### 18.1 LESSONS LEARNED

I'd like to conclude by revisiting the questions raised at the beginning of this book:

- To what degree has ecological forecasting emerged as a robust subdiscipline?
- Is the supply of ecological forecasts keeping up with the demand?
- Is ecology becoming a more predictive science?

For the first two points, as a community we are unfortunately not there *yet*. While ecology has been making predictions and informing policy and management for decades, this has not yet congealed into a community that is looking broadly across subdisciplines to develop the required theory, methods, and student training to meet these challenges. However, interest in this area is emerging rapidly and I hope this book has brought you, the reader, further along that road. What I see as the frontiers for this nascent field is the topic of section 18.2.

That said, for the third point the answer is a definite "yes." Looking across the case studies (chapters 7, 10, 12, and 15), the capacity to produce forecasts with well-specified uncertainties exists across many ecological subdisciplines. However, different fields have different strengths, and to the best of my knowledge all the topics covered in this book have yet to come together in one place. Figure 18.1 revisits the overall workflow from chapter 1 to illustrate what putting the concepts covered in this book into practice might look like.

Starting from the beginning (left-hand side), ecological forecasting is fundamentally a data-driven enterprise. The rapid growth in available data has led to an increased need to employ formal informatics approaches to manage the flows of data into analyses (chapter 3) and to automate the processing of whole workflows (chapter 4). Likewise, there is a growing recognition of the need to adhere to best practices and ensure open, repeatable analyses. This not only leads to better science, but it also makes that science more robust to outside criticism. While I lack data to back this up, my casual observation is that there has been a general increase in the baseline computational skills of ecologists, particularly young ecologists, and that there is rapidly growing expertise in the area of ecoinformatics. Considerably more attention is being paid to these issues than ever before and the landscape of available tools and resources is evolving rapidly.

The other area where all ecological subdisciplines are advancing is in the use of data to calibrate models. Sophisticated Bayesian and Maximum Likelihood methods are routinely employed to fit models to data, with the primary output of such analyses

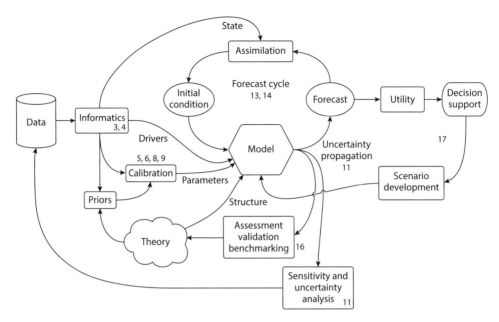

FIGURE 18.1. Ecological Forecasting conceptual workflow. Relevant chapters identified by number.

being data-constrained parameter estimates (chapter 5). That said, there's still a lot of variability in the extent to which different analyses deal with the complexities of real-world data (chapter 6). For example, some communities are adept at using hierarchical models to partition uncertainties into different sources, while in others such approaches are still rare. Across the board, most analyses still use Gaussian likelihoods and uninformative priors.

One of the largest challenges for all communities, though simultaneously one of the greatest opportunities, lies with combining data from different sources (chapter 9). Particularly challenging is dealing with data when there are either very unbalanced sample sizes (orders of magnitude more data for one type of measurement than for another) or when data cross different scales (for example, differences in timescale, spatial scale, or phylogenetic resolution). Statistical methods for combining data across scales exist, but become computationally challenging when the differences in scale are large. Furthermore, many ecological processes are scale dependent, and thus statistical downscaling or upscaling across too large a difference in scale also runs into theoretical challenges. Process-based models that include scale explicitly present an important alternative for combining information across scales. More generally, best practice for data fusion, whether unbalanced or not, requires that we address the complexities and information content of data (autocorrelation, sampling, observation errors, and so on), but unbalanced data requires that we pay extra attention to the systematic errors in the data and the model.

Given a well-calibrated model, the number-one factor that separates ecological forecasting from the simple forward simulation of models is the attention given to uncertainty propagation. A wide range of methods are available for uncertainty propagation (chapter 11), with the choice among them driven by trade-offs in computation and complexity between analytical and numerical methods and methods that produce just estimates of statistical moments (for example, mean, variance) and those that produce full PDFs. Regardless of the approach, all methods provide a way of propagating not just parameter uncertainty from the calibration, but also uncertainties in initial conditions, model drivers/inputs, and model structure.

In addition to the primary, forward flow of information from data into models and then into forecasts, our conceptual workflow also highlights three important feedback loops aimed at reducing uncertainties in model parameters, structure, and initial conditions.

The first of these is the use of sensitivity and uncertainty analyses to identify the dominant sources of uncertainty in model forecasts (chapter 11). These analyses feed back to data collection—identifying key uncertainties allows us to target these for further measurement and data synthesis, either informally or through more formal observation design analyses, such as Observing System Simulation Experiments (OSSEs). These methods can be applied to all model inputs (parameters, initial conditions, drivers, structure, and so on), though most of the ecological literature focuses on parameter uncertainty. One of the key take-home messages from this literature is that both model sensitivity and input (for example, parameter) uncertainty are equally important in identifying the dominant sources of uncertainty in model output—we can't rely on just one or the other alone when trying to refine data collection.

The second feedback loop in our workflow comes from model assessment (chapter 16). The analysis of model residual error and the validation of model outputs against new, independent data are both critical for identifying systematic errors in model structure. Uncertainty propagation is also important for model assessment—it is impossible to assess a model's structural errors if its confidence interval is so wide that all observations fall within it. Benchmarks form a special class of assessments, ones that are typically agreed upon by a community to allow a fair comparison among models, to track model improvement through time, and to set a minimum bar for model skill. When assessing model skill, it is also important to establish an appropriate and nontrivial null model that represents our prediction in the absence of a formal ecological forecast. Ultimately, model assessments are not meant to just make us feel good about how well our model is doing, but to identify deficiencies in current theories, test among competing hypotheses, and improve model structure.

The third feedback loop is the forecast cycle, which is used to refine estimates of the model's initial conditions. The forecast cycle is a special case of the more general forecasting problem, focused on combining data and models to constrain estimates of the state of the system (population size, biomass, and so on), since ultimately it is the state of the system that we're most often trying to forecast. In chapter 8 the state-space framework (also known as the Hidden Markov model) was introduced as the general case for this problem, wherein the state of the system is treated as a latent variable—a quantity that we estimate through data but never observe directly. The state-space framework can be very flexibly applied across both time and space and is actively used in a number of ecological research communities.

The forecast cycle aims to update forecasts when new data becomes available. This idea of updatability is central to forecasting, and is particularly essential in systems with either stochastic (for example, disturbance) or chaotic dynamics. The different variants of data assimilation used in the forecast cycle (chapters 13 and 14) reflect the different approaches used for uncertainty propagation (chapter 11): the Kalman filter is an analytical propagation of the mean and variance; the Extended Kalman filter is a linear approximation to these same moments; the Ensemble Kalman filter is a ensemble-based numeric approximation to these moments; and the particle filter is a Monte Carlo approximation to the full PDF. To date there's been relatively limited adoption of formal forecast updating in ecology (but see chapter 15), and this remains an important research frontier, not just in application but also for methods development (section 18.2.2).

Finally, a common goal of ecological forecasting is to support decision making (chapter 17). Forecasting plays a critical role in environmental decision making because fundamentally decisions are about what will happen in the future, in response to our choices, rather than about what's already happened in the past. Engagement with stakeholders and decision makers is critical to this process, to ensure that ecologists are forecasting the variables that are of interest and are reporting the performance measures needed. Ecologists are often quick to ask what threshold level of uncertainty they must get under for a forecast to be useful to decision makers (Petchey et al. 2015). In other words, how good is good enough? However, it is important to understand that decision makers are unlikely, a priori, to have a precise answer to that question. In decision making uncertainty translates into risk, and risk tolerance varies among individuals and with the nature of the decision—individuals tend to be more risk adverse under one-time, high-impact decisions than repeated, routine decisions. If there is a "threshold" toward which an ecological forecaster should aspire, it's the one where decisions become insensitive to (also known as robust to) the uncertainties. However, such thresholds typically emerge post hoc out of the policy analysis itself and will depend a lot on the set of alternatives considered. That means for most environmental decisions there will never be a clearly defined a priori bar that we need to hit in order to be useful. Indeed, the criteria that I've been told in multiple conversations is one ecologists should find encouraging and empowering— forecasts need only be better than chance (50:50) to be useful. That said, if a forecast is genuinely uncertain, then that uncertainty needs to be accurately portrayed; underreporting uncertainty leads to falsely confident decisions and that exposure to risk undermines a forecaster's credibility, while overreporting uncertainty leads to overly cautious decisions that may strain limited resources.

## 18.2 FUTURE DIRECTIONS

### 18.2.1 The Theory of Ecological Forecasting

As discussed in chapter 2, we as a community have only scratched the surface in defining the theoretical basis for ecological forecasting. The objective of this book was more focused on putting forecasts into practice, not to lay such a foundation, but I do think we've seen many of the important bricks along the way. The first of these are ideas about *inherent predictability*. While there are not fundamental equations of ecology to study, we know more generally that ecosystems are subject to numerous *feedbacks*. It is fair to say that positive, stabilizing feedbacks will tend to increase the inherent predictability of a system. On the other side, negative, destabilizing feedbacks will tend to decrease predictability, often very dramatically at the point where a system transitions into chaos. From this we see that these forces are not symmetric—a little bit of chaos can cause a lot of trouble (chapter 15). More subtly, if we look back at our definition of chaos (see box 2.2) as a sensitivity to initial conditions, as well as the long-taught stability analyses in population modeling, we see that they both presuppose an "initial conditions problem." This is the very problem that most forecasting methods have been designed to solve (for example, weather forecasting), but is it the problem that ecologists face? As discussed in chapters 2, 11, and 13, when ecological systems have stabilizing feedbacks, then first principles suggest that other sources of uncertainty, such as process error, will dominate over initial condition error. While there has not been a systematic study of ecological process error, case studies to date suggest it is nontrivial (Clark 2003; Clark and Bjørnstad 2004). First-principles analyses that include a larger suite of possible uncertainties (section 2.5.3) highlight the critical distinction between sources of uncertainty that grow or decline exponentially from those that grow linearly. Also critical are sources of uncertainty that decline asymptotically with additional data (for example, parameters) from those that are inherently irreducible (for example, stochasticity).

Consider again the example in chapter 6 of trying to forecast ten sites into the future (see figure 6.7). In one case there was large *site-to-site variability*, but these differences were relatively persistent from year to year, meaning that forecasting in time for an existing site could be done much more confidently than forecasting a new site. This example highlights the importance of *system memory, site history, and heterogeneity*, all of which are ubiquitous in ecological systems. These factors are also frequently cursed by both modelers and field ecologists alike as making it hard to generalize and extrapolate. But they are also very different from an initial condition problem. First, differences among sites are not "perturbations" that we expect to either converge back to the across-site mean or lead to chaotic fluctuations. Second, as we continue to observe any particular site, we learn more about it and become more confident in forecasting it. This is not true for initial condition problems, such as weather forecasting, and thus it is not accounted for in data assimilation approaches developed for initial condition problems. Furthermore, most traditional modeling approaches deal with system memory by introducing lags into models (for example, looking back over the previous *n* time steps, rather than just at the current system state, where *n* is typically small), which statistically is equivalent to shifting to an autoregressive model with *n* lags (Turchin 2003). However, for many ecological problems system memory is not operating over a few time steps, but operating on completely different spatial and temporal scales associated with fundamentally different processes (pedogenesis, evolution, tectonics, and so on). Over the timescales of most ecological time-series they are indistinguishable from persistent random effects. Indeed, a random effect can be thought of as a special case of autocorrelation with lag infinity.

Practically, however, while variability in ecological systems can be large, it isn't unbounded. A big unknown in many cases isn't just the presence of variability, but the spatial, temporal, and phylogenetic *scales of variability*. Differences in variability across scales can make predicting the same process very different depending upon what scale is examined. In general, as one goes up in scale the law of large numbers tends to make certain dynamics more predictable. For example, mortality and disturbance can be extremely stochastic at a fine scale, but very predictable at large scales when averaging over many such events (Turner et al. 1993). From the regional perspective looking back down-scale to the local, we may then find that while individual sites are not predictable, their overall distribution is. In many cases the ability to forecast that distribution accurately may be much more useful than predicting a specific location inaccurately. In some cases we can even use these ideas to derive explicit scaling relationships between local and regional processes (Moorcroft et al. 2001). However, as we move up in scale, some factors that previously appeared to be static or random begin to show systematic patterns and gradients (for example, climate). Depending on the balance between these tendencies, predictability may increase, decrease, or show multiple minima and maxima as we traverse scales. Overall, a key area of research in forecasting is to better understand when and where these scales of predictability are in different systems and where those do and don't align with our forecasting needs.

Closely related to the issues of scale and feedbacks is the idea of *emergent phenomena*. For example, in a plant community the total leaf area index (LAI) operates within a relatively narrow band that is fairly predictable based on climate and stand age. By contrast, the LAI of an individual species in the system can be highly variable due to responses to finer scale environmental variables (for example, soils) and a good bit of stochastic chance (for example, dispersal, mortality). Predicting the LAI of the stand by summing up the LAI predictions of each species would likely be biased and highly uncertain. This emergent behavior of a predictable outcome occurring despite unpredictable parts arises because of stabilizing feedbacks (for example, light and resource competition) that are stronger at the ecosystem level than at the individual level. Statistically, this is equivalent to saying that there are strong covariances among the parts of the system.

Ecologically, the concept of emergent phenomena—that the whole is more than the sum of its parts—is foundational to our science, but in practice this idea does not always penetrate into the way we make forecasts. In particular, there's been a strong conviction in the modeling community that we should independently calibrate the different processes in models then test whether we can predict the whole. Undoubtedly, detailed process-level studies remain necessary to constrain the parts and ensure that we don't get the right answer for the wrong reason (LeBauer et al. 2013). But building models piecemeal means that we are unable to estimate the covariances among the different parts—if those tend to be negative (stabilizing), then we end up with potential biases and overestimate the uncertainty in emergent phenomena (Clark et al. 2014). Capturing the covariances among these parts requires not only that we fit the whole model at once, but that we also bring together data from across scales—that we combine data about both the parts and the whole. This remains a complex problem in data fusion with many opportunities for further research (chapters 9 and 10).

Overall, what we're seeing from these examples is that as we continue to develop the theoretical basis of forecasting—building on foundational concepts such as inherent predictability, variability, scale-dependence, and emergent phenomena—there is likely to be a tight link to methods developments. Rather than drawing a distinction between process and statistical models, our process models need to be treated with a good deal of statistical sophistication. If ecological systems are not an "initial condition problem," but rather a "random effects problem" or an "all of the above" problem (drivers, parameters, initial conditions, structure), then existing data assimilation approaches borrowed from other disciplines are inadequate and new tools need to be developed and explored. In doing so the two maxims of "Know thy data!"

and "Know thy system!" are critical guiding principles for any ecologist and are not going away any time soon. However, I remain confident that the emerging field of ecological forecasting will develop approaches and concepts that will be broadly applicable across subdisciplines, and that much is to be gained from the cross-fertilization of looking at the forecasting problem across the field of ecology. A forecasting perspective has the potential to not only make ecology much more relevant in the face of numerous environmental challenges, but to transform the basic science and theory of the field.

### 18.2.2 Toward Real-Time Ecological Forecasts

I began this book discussing Clark et al.'s (2001) rallying cry that ecology needed to become a more predictive science. I'd like to end by throwing down a gauntlet of my own:

> Ecological forecasts need to be closer to real-time.

The motivation for such an undertaking is that while there are lots of long-term forecasts being produced (for example, ecological impacts of climate change on 100 year horizons), many ecological problems (and most decisions) are occurring over a much shorter term. Short-term forecasts need the ability to respond to the real world as it evolves in response to disturbance and extreme events, and as new information becomes available, ideally in an automated or semi-automated manner. The short-term decisions made by managers—which are the vast majority of management decisions—cannot wait for the academic publication cycle. Humankind fundamentally cares about the world around it, and there is an enormous opportunity in providing real-time information about that natural world. From remote sensing, to sensor and monitoring networks, to long-term research, we as ecologists have invested enormous time and effort into kick-starting an era of ecological big data. Iterative forecasting (chapters 13 and 14) is the natural arc of that effort. If history is any judge, we will be amazed by the innovative ways that society will put such forecasts to use.

In addition to being more socially relevant, iterative forecasting has the potential to transform the science of ecology. If we look at the field of meteorology, the first forecasts made by process-based models were decidedly unimpressive. It took years of effort from the forecasting community to get the point where these models performed better than simple statistical extrapolations from current conditions and past events. However, not only did forecasts improve beyond that point, but also their skill has continued to improve almost linearly over the past 60 years. This improvement came from steady improvements in models, in data, in assimilation techniques, and in computing. All but the last of these were internal to the discipline and very much driven by the daily feedback provided by the process of making real-time forecasts. In essence, the atmospheric science community got to test a hypothesis (that day's forecast) every single day, year after year, decade after decade, over 20,000 times and counting. That kind of feedback is simply absent from ecology. There are the two critical lessons from this experience: first, that we will likely stink at forecasting when we first get started, but that's OK so long as we are honest about our uncertainties; and second, that the only way to get better is to start trying. We cannot sit back and wait until our models are better and our data are better to start forecasting because if we do that they will never be good enough. There will always be an argument to wait until we understand more, and measure more, and model more, but society is not going to wait. The missed opportunities are going to be real and come at the cost of both natural systems and human welfare.

Now is the time to jump on the learning curve.

## References

### 1. INTRODUCTION

Clark, J. S. 2005. "Why Environmental Scientists Are Becoming Bayesians." *Ecol. Lett.*, 8, 2–14.

———. 2007. *Models for Ecological Data: An Introduction.* Princeton, NJ: Princeton University Press.

Clark, J. S., S. R. Carpenter, M. Barber, S. Collins, A. Dobson, J. A. Foley, et al. 2001. "Ecological Forecasts: An Emerging Imperative." *Science*, 293, 657–60.

Dietze, M. C., D. S. LeBauer, and R. Kooper. 2013. "On Improving the Communication between Models and Data." *Plant. Cell Environ.*, 36, 1575–85.

Dietze, M. C., S.P.S. Serbin, C. D. Davidson, A. R. Desai, X. Feng, R. Kelly et al. 2014. "A Quantitative Assessment of a Terrestrial Biosphere Model's Data Needs across North American Biomes." *J. Geophys. Res. Biogeosci.*, 119 (3), 286–300.

Ellison, A. M. 2004. "Bayesian Inference in Ecology." *Ecol. Lett.*, 7, 509–20.

Hobbs, N. T., and M. B. Hooten 2015. *Bayesian Models: A Statistical Primer for Ecologists.* Princeton, NJ: Princeton University Press.

Kahneman, D. 2013. *Thinking Fast and Slow.* New York: Farrar, Straus and Giroux.

LeBauer, D., D. Wang, K. T. Richter, C. C. Davidson, and M. C. Dietze 2013. "Facilitating Feedbacks between Field Measurements and Ecosystem Models." *Ecol. Monogr.*, 83, 133–54.

Medlyn, B. E., M. G. De Kauwe, A. P. Walker, M. C. Dietze, P. Hanson, T. Hickler et al. 2015. "Using Ecosystem Experiments to Improve Vegetation Models: Lessons Learnt from the Free-Air $CO_2$ Enrichment Model-Data Synthesis." *Nat. Clim. Chang.*, 5, 528–34.

United Nations. 2014. *Concise Report on the World Population Situation in 2014.* Department of Economic and Social Affairs, Population Division, ST/ESA/SER.A/354. New York: United Nations.

Walker, A. P., P. J. Hanson, M. G. De Kauwe, B. E. Medlyn, S. Zaehle, S. Asao *et al.* 2014. "Comprehensive Ecosystem Model Data Synthesis Using Multiple Data Sets at Two Temperate Forest Free Air $CO_2$ Enrichment Experiments: Model Performance at Ambient $CO_2$ Concentration." *J. Geophys. Res.*, 119, 937–64.

Walters, C. 1986. *Adaptive Management of Renewable Resources.* New York: Macmillan.

### 2. FROM MODELS TO FORECASTS

Armstrong, J. 2001. *Principles of Forecasting: A Handbook for Researchers and Practitioners.* New York: Kluwer Academic.

Beckage, B., L. J. Gross, and S. Kauffman. 2011. "The Limits to Prediction in Ecological Systems." *Ecosphere*, 2, art125.

Bjørnstad, O. N., B. F. Finkenstadt, and B. T. Grenfell. 2002. "Dynamics of Measles Epidemics: Estimating Scaling of Transmission Rates Using a Time Series SIR Model. *Ecol. Monogr.*, 72, 169–84.

Davis, M. B. 1983. "Quaternary History of Deciduous Forests of Eastern North America and Europe." *An. Mo. Bot. Gar.*, 70, 550–63.

Ellner, S., and P. Turchin. 1995. "Chaos in a Noisy World: New Methods and Evidence from Time-Series Analysis." *Am. Nat.*, 145, 343–75.

Gelfand, A. E., and S. K. Ghosh. 1998. "Model Choice: A Minimum Posterior Predictive Loss Approach." *Biometrika*, 85, 1–11.

Hawkins, E., and R. Sutton. 2009. "The Potential to Narrow Uncertainty in Regional Climate Predictions." *Bull. Am. Meteorol. Soc.*, 90, 1095–1107.

Hoeting, J. A., D. Madigan, A. E. Raftery, and C. T. Volinsky. 1999. "Bayesian Model Averaging: A Tutorial." *Statistical Science*, 14, 382–417.

Kalnay, E. 2002. *Atmospheric Modeling, Data Assimilation and Predictability*. Cambridge: Cambridge University Press.

Lande, R. 1993. "Risks of Population Extinction from Demographic and Environmental Stochasticity and Random Catastrophes." *Am. Nat.*, 142, 911–27.

May, R. 1976. "Simple Mathematical Models with Very Complicated Dynamics." *Nature*, 261, 459–67.

Otto, S. P., and T. Day. 2007. *A Biologist's Guide to Mathematical Modeling in Ecology and Evolution*. Princeton, NJ: Princeton University Press.

Press, W. H., S. A. Teukolsky, W. T. Vettering, and B. P. Flannery. 2007. *Numerical Recipes: The Art of Scientific Computing*. 3rd ed. Cambridge: Cambridge University Press.

Silver, N. 2012. *The Signal and the Noise*. London: Penguin Press.

Taleb, Nassim Nicholas. 2007. *The Black Swan: The Impact of the Highly Improbable*. 2nd ed. New York: Random House.

Turner, M., W. Romme, and R. Gardner. 1993. "A Revised Concept of Landscape Equilibrium: Disturbance and Stability on Scaled Landscapes." *Landscape Ecol.*, 8, 213–27.

US Department of Defense. 2002. "DoD News Briefing—Secretary Rumsfeld and General Myers." Available at http://archive.defense.gov/Transcripts/Transcript.aspx?TranscriptID =2636.

## 3. DATA, LARGE AND SMALL

Arge, L., J. S. Chase, P. Halpin, L. Toma, V. S. Vitter, D. Urban et al. 2003. "Efficient Flow Computation on Massive Grid Terrain Datasets." *Geoinformatica*, 7, 283–313.

Beardsley, T. M. 2014. "Are You Ready for Open Data?" *Bioscience*, 64, 855–55.

DAMA International. 2010. *The DAMA Guide to the Data Management Body of Knowledge (DAMA-DMBOK)*. Denville, NJ: Technics Publications.

Dean, J., and S. Ghemawat. 2008. "MapReduce: Simplified Data Processing on Large Clusters." *Commun. ACM*, 51 (1), 107–11.

Dietze, M. C., and P. R. Moorcroft. 2011. "Tree Mortality in the Eastern and Central United States: Patterns and Drivers." *Glob. Chang. Biol.*, 17, 3312–26.

Goodman, A., A. Pepe, A. W. Blocker, C. L. Borgman, K. Cranmer, M. Crosas et al. 2014. "Ten Simple Rules for the Care and Feeding of Scientific Data." *PLoS Comput. Biol.*, 10, e1003542.

Govindarajan, S., M. C. Dietze, P. K. Agarwal, and J. S. Clark. 2007. "A Scalable Algorithm for Dispersing Population." *J. Intell. Inf. Syst.*, 29, 39–61.

Hampton, S. E., S. S. Anderson, S. C. Bagby, C. Gries, X. Han, E. M. Hart et al. 2014. "The Tao of Open Science for Ecology." *Ecosphere*, 6 (7), art120.

Hampton, S. E., C. A. Strasser, J. J. Tewksbury, W. K. Gram, A. E. Budden, A. L. Batcheller et al. 2013. "Big Data and the Future of Ecology." *Front. Ecol. Environ.*, 11 (3), 156–62.

Jones, M. B., M. P. Schildhauer, O. J. Reichman, and S. Bowers. 2006. "The New Bioinformatics: Integrating Ecological Data from the Gene to the Biosphere." *Annu. Rev. Ecol. Evol. Syst.*, 37, 519–44.

Michener, W. K., and M. B. Jones. 2012. "Ecoinformatics: Supporting Ecology as a Data-intensive Science." *Trends Ecol. Evol.*, 27, 85–93.

Palmer, C. L., M. H. Cragin, P. B. Heidorn, and L. C. Smith. 2007. "Data Curation for the Long Tail of Science: The Case of Environmental Sciences." In *Digital Curation Conference*, Washington, DC.

Reichman, O. J., M. B. Jones, and M. P. Schildhauer. 2011. "Challenges and Opportunities of Open Data in Ecology." *Science*, 331, 703–5.

Schimel, D. S. (2016). "Editorial." *Ecol. Appl.*, 26, 3–4.

Strasser, C., R. Cook, W. Michener, and A. Budden. 2012. *Primer on Data Management: What You Always Wanted to Know*. Available at www.dataone.org/sites/all/documents/DataONE _BP_Primer_020212.pdf.

Vitolo, C., Y. Elkhatib, D. Reusser, C.J.A. Macleod, and W. Buytaert. 2015. "Web Technologies for Environmental Big Data." *Environ. Model. Softw.*, 63, 185–98.

Vitter, J. S. (2001). "External Memory Algorithms and Data Structures: Dealing with Massive Data." *ACM Comput. Surv.*, 33, 209–71.

White, E., E. Baldridge, Z. Brym, K. Locey, D. McGlinn, and S. Supp. 2013. "Nine Simple Ways to Make It Easier to (Re)use Your Data." *Ideas Ecol. Evol.*, 6, 1–10.

## 4. SCIENTIFIC WORKFLOWS AND THE INFORMATICS OF MODEL-DATA FUSION

Boose, E. R., A. M. Ellison, L. J. Osterweil, L. A. Clarke, R. Podorozhny, J. L. Hadley et al. 2007. "Ensuring Reliable Datasets for Environmental Models and Forecasts." *Ecol. Inform.*, 2, 237–47.

Ellison, A. M. 2010. "Repeatability and Transparency in Ecological Research." *Ecology*, 91, 2536–39.

Ellison, A. M., L. Osterweil, L. Clarke, and J. Hadley. 2006. "Analytic Webs Support the Synthesis of Ecological Data Sets." *Ecology*, 87, 1345–58.

Goble, C., and D. De Roure. 2009. "The Impact of Workflow Tools on Data-centric Research." In *The Fourth Paradigm: Data-Intensive Scientific Discovery.*, ed. T. Hey, S. Tansley, and K. Tolle. Redmond, WA: Microsoft Research, 137–45.

Jacoby, G. C., and R. D. D'Arrigo. 1995. "Tree Ring Width and Density Evidence of Climatic and Potential Forest Change in Alaska." *Global Biogeochem. Cycles*, 9, 227–34.

Ludäscher, B., I. Altintas, C. Berkley, D. Higgins, E. Jaeger, M. Jones et al. 2006. "Scientific Workflow Management and the Kepler System." *Concurr. Comput. Pract. Exp.—Work. Grid Syst.*, 18, 1039–65.

Sandberg, S. 2013. *Lean In: Women, Work, and the Will to Lead*. New York: Knopf.

Vines, T. H., A.Y.K. Albert, R. L. Andrew, F. Débarre, D. G. Bock, M. T. Franklin et al. 2014. "The Availability of Research Data Declines Rapidly with Article Age." *Curr. Biol.*, 24, 94–97.

Wilson, G., D. A. Aruliah, C. T. Brown, N. P. Chue Hong, M. Davis, R. T. Guy et al. 2014. "Best Practices for Scientific Computing." *PLoS Biol.*, 12 (1), e1001745.

## 5. INTRODUCTION TO BAYES

Ayyub, B. M. 2000. *Methods for Expert-Opinion Elicitation of Probabilities and Consequences for Corps Facilities*. Alexandria, VA: U.S. Army Corp of Engineers, Institute for Water Resources.

Boring, R., D. Gertman, J. Joe, and J. Marble. 2005. *Simplified Expert Elicitation Guideline for Risk Assessment of Operating Events*. Report INL/EXT-05-00433. Idaho Falls: Idaho National Laboratory.

Clark, J. S. 2007. *Models for Ecological Data: An Introduction*. Princeton, NJ: Princeton University Press.

EPA. 2009. *U.S. Environmental Protection Agency Expert Elicitation Task Force White Paper*. Washington, DC: U.S. Environmental Protection Agency, Science Policy Council.

Gelman, A., J. B. Carlin, H. S. Stern, D. B. Dunson, A. Vehtari, and D. B. Rubin. 2013. *Bayesian Data Analysis*. 3rd ed. Boca Raton, FL: Chapman & Hall/CRC Texts in Statistical Science (Book 106).

Gilks, W. R., and P. Wild. 1992. "Adaptive Rejection Sampling for Gibbs Sampling." *Appl. Stat.*, 41, 337–48.

Hilborn, R., and M. Mangel. 1997. *The Ecological Detective: Confronting Models with Data*. Princeton, NJ: Princeton University Press.

Hobbs, N. T., and M. B. Hooten. (2015). *Bayesian Models: A Statistical Primer for Ecologists*. Princeton, NJ: Princeton University Press.

Kahneman, D. 2013. *Thinking Fast and Slow*. New York: Farrar, Straus and Giroux.

Koricheva, J., J. Gurevitch, and K. Mengersen. 2013. *Handbook of Meta-analysis in Ecology and Evolution*. Princeton, NJ: Princeton University Press.

LeBauer, D., D, Wang, K. T. Richter, C. C. Davidson, and M. C. Dietze. 2013. "Facilitating Feedbacks between Field Measurements and Ecosystem Models. *Ecol. Monogr.*, 83, 133–54.

Link, W. A., and M. J. Eaton. 2012. "On Thinning of Chains in MCMC." *Methods Ecol. Evol.*, 3, 112–15.

Morgan, M. G. 2014. "Use (and Abuse) of Expert Elicitation in Support of Decision Making for Public Policy." *Proc. Natl. Acad. Sci. U.S.A.*, 111, 7176–84.

Plummer, M., N. Best, K. Cowles, and K. Vines. 2006. "CODA: Convergence Diagnosis and Output Analysis for MCMC." *R News*, 6, 7–11.

Slottje, P., J. P. Van Der Sluijs, and A. Knol, A. 2008. *Expert Elicitation: Methodological Suggestions for Its Use in Environmental Health Impact Assessments*. Netherlands National Institute for Public Health and the Environment, RIVM Letter Report 630004001/2008. Available at http://www.nusap.net/downloads/reports/Expert_Elicitation.pdf.

## 6. CHARACTERIZING UNCERTAINTY

Banerjee, S., B. P. Carlin, and A. E. Gelfand. 2003. *Hierarchical Modeling and Analysis for Spatial Data*. Boca Raton, FL: Chapman & Hall/CRC Monographs on Statistics & Applied Probability.

Clark, J. S., M. C. Dietze, S. Chakraborty, P. K. Agarwal, M. S. Wolosin, I. Ibanez, and S. LaDeau. 2007. "Resolving the Biodiversity Paradox." *Ecol. Lett.*, 10, 647–59; discussion 659–62.

Cressie, N., and C. K. Wikle. 2011. *Statistics for Spatio-Temporal Data*. New York: Wiley.

Dietze, M. C., M. S. Wolosin, and J. S. Clark. 2008. "Capturing Diversity and Interspecific Variability in Allometries: A Hierarchical Approach." *Forest Ecol. Man.*, 256, 1939–48.

Diggle, P. J. 1990. *Time Series: A Biostatistical Introduction*. Oxford: Oxford University Press.

Gelman, A., J. B. Carlin, H. S. Stern, D. B. Dunson, A. Vehtari, and D. B. Rubin. 2013. *Bayesian Data Analysis*. 3rd ed. Boca Raton, FL: Chapman & Hall/CRC Texts in Statistical Science (Book 106).

Messinger, F., G. Dimego, E. Kalnay, K. Mitchell, and P. C. Shafran. 2006. "North American Regional Reanalysis." *Bull. Am. Meteorol. Soc.*, 87, 343–60.

Moffat, A., D. Papale, M. Reichstein, D. Hollinger, A. D. Richardson, A. Barr, C. Beckstein, B. Braswell, G. Churkina, and A. Desai. 2007. "Comprehensive Comparison of Gap-filling Techniques for Eddy Covariance Net Carbon Fluxes." *Agricult. Forest Meteorol.*, 147, 209–32.

O'Hara, R. B., and D. J. Kotze. 2010. "Do Not Log-transform Count Data." *Methods Ecol. Evol.*, 1, 118–22.

Spiegelhalter, D. J., N. G. Best, B. P. Carlin, and A. van der Linde. 2002. "Bayesian Measures of Model Complexity and Fit." *J. R. Stat. Soc. B*, 64, 583–639.

Stoppard, T. 1994. *Rosencrantz and Guildenstern Are Dead*. Reprint ed. New York: Grove Press.

Townshend, J. R., J. G. Masek, C. Huang, E. F. Vermote, F. Gao, S. Channan et al. 2012. "Global Characterization and Monitoring of Forest Cover Using Landsat Data: Opportunities and Challenges." *Int. J. Digital Earth*, 5, 373–97.

Wenger, S. J., and M. C. Freeman. 2008. "Estimating Species Occurrence, Abundance, and Detection Probability Using Zero-inflated Distributions." *Ecology*, 89, 2953–59.

## ▬ 7. CASE STUDY: BIODIVERSITY, POPULATIONS, AND ENDANGERED SPECIES

Birch, L. C. 1948. "The Intrinsic Rate of Natural Increase of an Insect Population." *J. Anim. Ecol.*, 17, 15–26.

Botkin, D. B., H. Saxe, M. B. Araújo, R. Betts, R.H.W. Bradshaw, T. Cedhagen et al. 2010. "Forecasting the Effects of Global Warming on Biodiversity." *BioScience* 57, 227–36.

Bugmann, H. 2001. "A Review of Forest Gap Models." *Clim. Change*, 51, 259–305.

Caswell, H. 2001. *Matrix Population Models*. 2nd ed. Sunderland, MA: Sinauer Associates.

Clark, J. S., A. E. Gelfand, C. W. Woodall, and K. Zhu. 2014. "More Than the Sum of the Parts: Forest Climate Response from Joint Species Distribution Models." *Ecol. Appl.*, 24, 990–99.

Cramer, W., A. Bondeau, F. I. Woodward, I. C. Prentice, R. A. Betts, V. Brovkin et al. 2001. "Global Response of Terrestrial Ecosystem Structure and Function to $CO_2$ and Climate Change: Results from Six Dynamic Global Vegetation Models." *Glob. Chang. Biol.*, 7, 357–73.

Crouse, D. T., and L. B. Crowder. 1987. "A Stage-Based Population Model for Loggerhead Sea Turtles and Implications for Conservation." *Ecol. Soc. Am.*, 68, 1412–23.

Crowder, L. B., D. T. Crouse, S. S. Heppell, and T. H. Martin. 1994. "Predicting the Impact of Turtle Excluder Devices on Loggerhead Sea Turtle Populations." *Ecol. Appl.*, 4 (3), 437–445.

de Valpine, P., K. Scranton, J. Knape, K. Ram, and N. J. Mills. 2014. "The Importance of Individual Developmental Variation in Stage-structured Population Models." *Ecol. Lett.*, 1026–38.

Ellner, S. P., and M. Rees. 2006. "Integral Projection Models for Species with Complex Demography." *Am. Nat.*, 167, 410–28.

———. 2007. "Stochastic Stable Population Growth in Integral Projection Models: Theory and Application. *J. Math. Biol.*, 54, 227–56.

Evans, M., K. Holsinger, and E. Menges. 2010. "Fire, Vital Rates, and Population Viability: A Hierarchical Bayesian Analysis of the Endangered Florida Scrub Mint." *Ecol. Monogr.*, 80, 627–49.

Evans, M., C. Merow, S. Record, S. M. McMahon, and B. J. Enquist. 2016. "Towards Process-based Range Modeling of Many Species." *Trends Ecol. Evol.*, 31 (11), 860–71.

Fox, G. A., and J. Gurevitch. 2000. "Population Numbers Count: Tools for Near Term Demographic Analysis." *Am. Nat.*, 156, 242–56.

Ibáñez, I., J. S. Clark, M. C. Dietze, M. H. Hersh, K. Feeley, S. LaDeau et al. 2006. "Predicting Biodiversity Change: Outside the Climate Envelope, beyond the Species-area Curve." *Ecology*, 87, 1896–1906.

Lande, R. 1988. "Demographic Models of the Northern Spotted Owl (*Strix occidentalis caurina*)." *Oecologia*, 75, 601–7.

Leslie, P. H. 1945. "On the Use of Matrices in Certain Population Mathematics." *Biometrika*, 33, 183–212.

Lotka, A. J. 1910. "Contribution to the Theory of Periodic Reaction." *J. Phys. Chem*, 14, 271–74.

Merow, C., J. P. Dahlgren, C.J.E. Metcalf, D. Z. Childs, M.E.K. Evans, E. Jongejans et al. 2014. "Advancing Population Ecology with Integral Projection Models: A Practical Guide." *Methods Ecol. Evol.*, 5, 99–110.

Morris, W., D. Doak, M. Groom, P. Kareiva, J. Fieberg, L. Gerber et al. 1999. *A Practical Handbook for Population Viability Analysis*. Arlington, VA: Nature Conservancy.

Morris, W. F., and D. F. Doak. 2002. *Quantitative Conservation Biology: Theory and Practice of Population Viability Analysis*. Sunderland, MA: Sinauer Associates.

Pacala, S. W., C. D. Canham, J. Saponara, J. A. Silander Jr., R. K. Kobe, and E. Ribbens. 1996. "Forest Models Defined by Field Measurements: Estimation, Error Analysis and Dynamics." *Ecol. Monogr.*, 66, 1–43.

Preston, F. W. 1962. "The Canonical Distribution of Commonness and Rarity: Part I." *Ecology*, 43, 185–215.

Tilman, D. 1982. *Resource Competition and Community Structure (MPB-17)*. Princeton, NJ: Princeton University Press.

Turchin, P. 2003. *Complex Population Dynamics: A Theoretical/Empirical Synthesis (MPB-35)*. Princeton, NJ: Princeton University Press.

Volterra, V. 1926). "Variazioni e fluttuazioni del numero d'individui in specie animali conviventi." *Mem. Acad. Lincei Roma*, 2, 31–113.

Williams, J. W., H. M. Kharouba, S. Veloz, M. Vellend, J. Mclachlan, Z. Liu et al. 2013. "The Ice Age Ecologist: Testing Methods for Reserve Prioritization during the Last Global Warming." *Glob. Ecol. Biogeogr.*, 22, 289–301.

## 8. LATENT VARIABLES AND STATE-SPACE MODELS

Banerjee, S., B. P. Carlin, and A. E. Gelfand. 2003. *Hierarchical Modeling and Analysis for Spatial Data*. Boca Raton, FL: Chapman & Hall/CRC Monographs on Statistics & Applied Probability.

Cressie, N., and C. K. Wikle. 2011. *Statistics for Spatio-Temporal Data*. New York: Wiley.

Jacquemoud, S., F. Baret, B. Andrieu, F. M. Danson and K. Jaggard. 1995. "Extraction of Vegetation Biophysical Parameters by Inversion of the PROSPECT + SAIL Models on Sugar Beet Canopy Reflectance Data. Application to TM and AVIRIS Sensors." *Remote Sens. Environ.*, 52, 163–72.

Korb, K. B., and A. E. Nicholson. 2011. *Bayesian Artificial Intelligence*. 2nd ed. Boca Raton, FL: CRC Press, Taylor & Francis Group.

LaDeau, S. L., and J. S. Clark. 2006. "Elevated $CO_2$ and Tree Fecundity: The Role of Tree Size, Interannual Variability, and Population Heterogeneity." *Glob. Chang. Biol.*, 12, 822–33.

Speer, J. H. 2010. *Fundamentals of Tree-ring Research*. Tucson: University of Arizona Press.

Whitlock, R., and M. McAllister. 2009. "A Bayesian Mark-Recapture Model for Multiple-Recapture Data in a Catch-and-Release Fishery." *Can. J. Fish. Aquat. Sci.*, 66, 1554–68.

Wright, H. E., J. E. Kutzbach, T. Webb, W. F. Ruddiman, S.-P.F. Alayne, and P. J. Bartlein. 1993. *Global Climates since the Last Glacial Maximum*. Minneapolis: University of Minnesota Press.

## 9. FUSING DATA SOURCES

Banerjee, S., B. P. Carlin, and A. E. Gelfand. 2003. *Hierarchical Modeling and Analysis for Spatial Data*. Boca Raton, FL: Chapman & Hall/CRC Monographs on Statistics & Applied Probability.

Curtis, P. S., and X. Z. Wang. 1998. "A Meta-analysis of Elevated $CO_2$ Effects on Woody Plant Mass, Form and Physiology." *Oecologia*, 113, 299–313.

Keenan, T. F., E. A. Davidson, J. W. Munger, and A. D. Richardson. 2013. "Rate My Data: Quantifying the Value of Ecological Data for the Development of Models of the Terrestrial Carbon Cycle." *Ecol. Appl.*, 23, 273–86.

Koricheva, J., J. Gurevitch, and K. Mengersen. 2013. *Handbook of Meta-analysis in Ecology and Evolution*. Princeton, NJ: Princeton University Press.

Lau, J., E. M. Antman, J. Jimenez-Silva, B. Kupelnick, F. Mosteller, and T. C. Chalmers. 1992. "Cumulative Meta-analysis of Therapeutic Trials for Myocardial Infarction." *New Eng. J. Med.*, 327, 248–54.

Lau, J., C. H. Schmid, and T. C. Chalmers. 1995. "Cumulative Meta-analysis of Clinical Trials Builds Evidence for Exemplary Medical Care." *J. Clin. Epidemiol.*, 48, 45–57.

LeBauer, D., D. Wang, K. T. Richter, C. C. Davidson, and M. C. Dietze. 2013. "Facilitating Feedbacks between Field Measurements and Ecosystem Models." *Ecol. Monogr.*, 83, 133–54.

Medvigy, D. M., S. C. Wofsy, J. W. Munger, D. Y. Hollinger, and P. R. Moorcroft. 2009. "Mechanistic Scaling of Ecosystem Function and Dynamics in Space and Time: Ecosystem Demography Model Version 2." *J. Geophys. Res.*, 114, 1–21.

Munger, J. W., J. B. McManus, D. D. Nelson, M. S. Zahniser, E. A. Davidson, S. C. Wofsy et al. 2016. "Seasonality of Temperate Forest Photosynthesis and Daytime Respiration." *Nature*, 534, 680–83.

Phillips, C. L., N. Nickerson, D. Risk, and B. J. Bond. 2011. "Interpreting Diel Hysteresis between Soil Respiration and Temperature." *Glob. Change Biol.*, 17, 515–27.

Reichstein, M., E. Falge, D. Baldocchi, D. Papale, M. Aubinet, P. Berbigier et al. 2005. "On the Separation of Net Ecosystem Exchange into Assimilation and Ecosystem Respiration: Review and Improved Algorithm." *Glob. Change Biol.*, 11, 1424–39.

Richardson, A. D., M. Williams, D. Y. Hollinger, D.J.P. Moore, D. B. Dail, E. A. Davidson et al. 2010. "Estimating Parameters of a Forest Ecosystem C Model with Measurements of Stocks and Fluxes as Joint Constraints." *Oecologia*, 164, 25–40.

Stoy, P. C., and T. Quaife. 2015. "Probabilistic Downscaling of Remote Sensing Data with Applications for Multi-Scale Biogeochemical Flux Modeling." *PLoS One* 10, e0128935.

Williams, M., A. D. Richardson, M. Reichstein, P. C. Stoy, P. Peylin, H. Verbeeck et al. 2009. "Improving Land Surface Models with FLUXNET Data." *Biogeosciences*, 6, 1341–59.

## 10. CASE STUDY: NATURAL RESOURCES

Beverton, R.J.H., and S. J. Holt. 1957. *On the Dynamics of Exploited Fish Populations*. Caldwell, NJ: Blackburn Press.

Blackman, V. H. 1919. "The Compound Interest Law and Plant Growth." *Ann. Bot.*, 33, 353–60.

Bugmann, H. (2001). "A Review of Forest Gap Models." *Clim. Change*, 51, 259–305.

CEC. 2009. *Green Paper. Reform of the Common Fisheries Policy*. Brussels: European Commission.

El-Sharkawy, M. A. 2011. "Overview: Early History of Crop Growth and Photosynthesis Modeling. *BioSystems*, 103, 205–11.

FAO. 2014. *State of the World Fisheries & Aquaculture Report*. Rome: UN FAO.

Hardin, G. 1968. "Tragedy of the Commons." *Science*, 162, 1243–48.

HELCOM and IBSFC. 1999. *Baltic Salmon Rivers—Status in the Late 1990s as Reported by the Countries in the Baltic Region. Technical Report*. Stockholm: Swedish National Board of Fisheries.

Hilborn, R., and M. Mangel. 1997. *The Ecological Detective: Confronting Models with Data*. Princeton, NJ: Princeton University Press.

Hilborn, R., and C. J. Walters. 1991. *Quantitative Fisheries Stock Assessment: Choice, Dynamics and Uncertainty*. New York: Springer.

Kareiva, P., M. Marvier, and M. McClure. 2000. "Recovery and Management Options for Spring/Summer Chinook Salmon in the Columbia River Basin." *Science*, 290, 977–79.

Kuikka, S., J. Vanhatalo, and H. Pulkkinen. 2014. "Experiences in Bayesian Inference in Baltic Salmon Management." *Statistical Sci.*, 29, 42–49.

Kurlansky, M. 2009. *The Last Fish Tale*. New York: Riverhead.

———. 2010. *Cod: A Biography of the Fish That Changed the World*. London: Penguin.

Mann, C. C., and M. L. Plummer. 2000. "Can Science Rescue Salmon?" *Science*, 289, 716–19.

Mäntyniemi, S., and A. Romakkaniemi. 2002. "Bayesian Mark-Recapture Estimation with an Application to a Salmonid Smolt Population." *Can. J. Fish. Aquat. Sci.*, 59, 1748–58.

Michielsens, C. G., S. Mäntyniemi, and P. J. Vuorinen. 2006a. "Estimation of Annual Mortality Rates Caused by Early Mortality Syndromes (EMS) and Their Impact on Salmonid Stock-Recruit Relationships." *Can. J. Fish. Aquat. Sci.*, 63, 1968–81.

Michielsens, C. G., and M. K. McAllister. 2004. "A Bayesian Hierarchical Analysis of Stock-Recruit Data: Quantifying Structural and Parameter Uncertainties." *Can. J. Fish. Aquat. Sci.*, 61, 1032–47.

Michielsens, C. G., M. K. McAllister, S. Kuikka, T. Pakarinen, L. Karlsson, A. Romakkaniemi et al. 2006b. "A Bayesian State-Space Mark-Recapture Model to Estimate Exploitation Rates in Mixed-Stock Fisheries." *Can. J. Fish. Aquat. Sci.*, 63, 321–34.

Michielsens, C.G.J., M. K. McAllister, S. Kuikka, S. Mäntyniemi, A. Romakkaniemi, T. Pakarinen et al. 2008. "Combining Multiple Bayesian Data Analyses in a Sequential Framework for Quantitative Fisheries Stock Assessment." *Can. J. Fish. Aquat. Sci.*, 65, 962–74.

Millennium Ecosystem Assessment Board. 2005. *Ecosystems and Human Well-being: Synthesis.* Washington, DC: Island Press.

Punt, A. E., and R. Hilborn. 1997. "Fisheries Stock Assessment and Decision Analysis: The Bayesian Approach." *Rev. Fish Biol. Fish.*, 7, 35–63.

Tetlock, P., and D. Gardner. 2015. *Superforecasting: The Art and Science of Prediction.* New York: Crown.

Uusitalo, L., S. Kuikka, and S. Romakkaniemi. 2005. "Estimation of Atlantic Salmon Smolt Carrying Capacity of Rivers Using Expert Knowledge. *ICES J. Mar. Sci.*, 62, 708–22.

Varis, O., and S. Kuikka. 1997. "Joint Use of Multiple Environmental Assessment Models by a Bayesian Meta-model: The Baltic Salmon Case. *Ecol. Modell.*, 102, 341–51.

## 11. PROPAGATING, ANALYZING, AND REDUCING UNCERTAINTY

Bastos, L., and A. O'Hagan. 2009. "Diagnostics for Gaussian Process Emulators." *Technometrics*, 51 (4), 425–38.

Cariboni, J., D. Gatelli, R. Liska, and A. Saltelli. 2007. "The Role of Sensitivity Analysis in Ecological Modelling." *Ecol. Modell.*, 203 (1–2), 167–82.

Casella, G., and R. L. Berger. 2001. *Statistical Inference.* 2nd ed. Pacific Grove, CA: Duxbury Press.

Charney, J., M. Halem, and R. Jastrow. 1969. "Use of Incomplete Historical Data to Infer the Present State of the Atmosphere." *J. Atmos. Sci.*, 26 (5), 1160–63.

Dietze, M. C., S.P.S. Serbin, C. D. Davidson, A. R. Desai, X. Feng, R. Kelly et al. 2014. "A Quantitative Assessment of a Terrestrial Biosphere Model's Data Needs across North American Biomes." *J. Geophys. Res. Biogeosci.*, 19 (3), 286–300.

Julier, S., J. Uhlmann, and H. F. Durrant-Whyte. 2000. "A New Method for the Nonlinear Transformation of Means and Covariances in Filters and Estimators." *IEEE Trans. Automat. Contr.*, 45 (3), 477–82.

Kennedy, M., C. Anderson, A. O'Hagan, M. Lomas, I. Woodward, and A. Heinemeyer. 2006 "Quantifying Uncertainty in the Biospheric Carbon Flux for England and Wales." *J. R. Stat. Soc. A*, 171 (1), 109–35.

Kennedy, M. C., and A. O'Hagan. 2001. "Bayesian Calibration of Computer Models." *J. R. Stat. Soc. B*, 63 (3), 425–64.

Leakey, A. D., K. A. Bishop, and E. A. Ainsworth. 2012, "A Multi-biome Gap in Understanding of Crop and Ecosystem Responses to Elevated $CO_2$." *Curr. Opin. Plant Biol.*, 15 (3), 228–36.

LeBauer, D., D. Wang, K. T. Richter, C. C. Davidson, and M. C. Dietze. 2013. "Facilitating Feedbacks between Field Measurements and Ecosystem Models." *Ecol. Monogr.*, 83 (2), 133–54.

Leuning, R. 1995. "A Critical Appraisal of a Combined Stomatal-Photosynthesis Model for C3 Plants." *Plant Cell Environ.*, 18 (4), 339–55.

Masutani, M., T. W. Schlatter, R. M. Errico, A. Stoffelen, W. Lahoz, J. S. Woollen, and G. D. Emmitt. 2010a. "Observing System Simulation Experiments." In *Data Assimilation*, ed. W. Lahoz, B. Khattatov, and R. Menard. Berlin/Heidelberg: Springer, 647–79.

Masutani, M., J. S. Woollen, S. J. Lord, G. D. Emmitt, T. J. Kleespies, S. A. Wood et al. 2010b. "Observing System Simulation Experiments at the National Centers for Environmental Prediction." *J. Geophys. Res. Atmos.*, 115(7), D07101.

Medvigy, D. M., S. C. Wofsy, J. W. Munger, D. Y. Hollinger, and P. R. Moorcroft. 2009. "Mechanistic Scaling of Ecosystem Function and Dynamics in Space and Time: Ecosystem Demography Model Version 2." *J. Geophys. Res.*, 114 (G1), 1–21.

Mishra, A. K., and T. N. Krishnamurti. 2011. "Observing System Simulation Experiment for Global Precipitation Mission." *Pure Appl. Geophys.*, 169 (3), 353–65.

Oakley, J., and A. O'Hagan. 2004 "Probabilistic Sensitivity Analysis of Complex Models: A Bayesian Approach." *J. R. Stat. Soc. B*, 66 (3), 751–69.

O'Hagan, A. 2006. "Bayesian Analysis of Computer Code Outputs: A Tutorial." *Reliab. Eng. Syst. Saf.*, 91 (10–11), 1290–1300.

O'Hagan, A., M. C. Kennedy, and J. E. Oakley. 1998. "Uncertainty Analysis and Other Inference Tools for Complex Computer Codes." In *Bayesian Statistics* 6, ed. J. M. Bernardo, J. O. Berger, A. P. Dawid, and A.F.M. Smith. Oxford: Oxford University Press, 503–24.

Press, W. H., S. A. Teukolsky, W. T. Vettering, and B. P. Flannery. 2007. *Numerical Recipes: The Art of Scientific Computing*. 3rd ed. Cambridge: Cambridge University Press.

Pujol, G., and B. Iooss. 2014. *sensitivity: Global Sensitivity Analysis of Model Outputs*. R package version 1.10. Available at http://CRAN.R-project.org/package=sensitivity.

Sacks, J., W. J. Welch, T. J. Mitchell, and H. P. Wynn. 1989. "Design and Analysis of Computer Experiments." *Stat. Sci.*, 4 (4), 409–23.

Saltelli, A., M. Ratto, T. Andres, F. Campolongo, J. Cariboni, D. Gatelli et al. 2008. *Global Sensitivity Analysis. The Primer Title*. New York: John Wiley & Sons.

Schimel, D., R. Pavlick, J. B. Fisher, G. P. Asner, S. Saatchi, P. Townsend et al. 2014. "Observing Terrestrial Ecosystems and the Carbon Cycle from Space." *Glob. Chang. Biol.*, 21 (5), 1762–76.

Schimel, D. S. 1995. "Terrestrial Biogeochemical Cycles: Global Estimates with Remote Sensing." *Remote Sens. Environ.*, 51(1), 49–56.

## 12. CASE STUDY: CARBON CYCLE

Clark, J. S., A. E. Gelfand, C. W. Woodall, and K. Zhu. 2014. "More Than the Sum of the Parts: Forest Climate Response from Joint Species Distribution Models. *Ecol. Appl.*, 24, 990–99.

Davidson, C. D. 2012. "The Modeled Effects of Fire on Carbon Balance and Vegetation Abundance in Alaskan Tundra." Master's thesis, University of Illinois, Urbana-Champaign.

De Kauwe, M. G., B. E. Medlyn, S. Zaehle, A. P. Walker, M. C. Dietze, T. Hickler et al. 2013. "Forest Water Use and Water Use Efficiency at Elevated $CO_2$: A Model-Data Intercomparison at Two Contrasting Temperate Forest FACE Sites." *Glob. Chang. Biol.*, 19, 1759–79.

Dietze, M. C. 2014. "Gaps in Knowledge and Data Driving Uncertainty in Models of Photosynthesis. *Photosynth. Res.*, 19, 3–14.

Dietze, M. C., LeBauer, D. S., and R. Kooper. 2013. "On Improving the Communication between Models and Data." *Plant. Cell Environ.*, 36, 1575–85.

Dietze, M. C., S.P.S. Serbin, C. D. Davidson, A. R. Desai, X. Feng, R. Kelly et al. 2014. "A Quantitative Assessment of a Terrestrial Biosphere Model's Data Needs across North American Biomes." *J. Geophys. Res. Biogeosci.*, 119 (3), 286–300.

Farquhar, G., S. Caemmerer, and J. A. Berry. 1980. "A Biochemical Model of Photosynthetic $CO_2$ Assimilation in Leaves of C3 Species." *Planta*, 149, 78–90.

Feng, X., and M. C. Dietze. 2013. "Scale Dependence in the Effects of Leaf Ecophysiological Traits on Photosynthesis: Bayesian Parameterization of Photosynthesis Models." *New Phytol.*, 200, 1132–44.

Fox, A., M, Williams, A. D. Richardson, D. Cameron, J. H. Gove, T. Quaife et al. 2009. "The REFLEX Project: Comparing Different Algorithms and Implementations for the Inversion of a Terrestrial Ecosystem Model against Eddy Covariance Data." *Agric. For. Meteorol.*, 149, 1597–1615.

Friedlingstein, P., P. Cox, R. Betts, L. Bopp, and W. Von. 2006. "Carbon-cycle Feedback Analysis: Results from the C4MIP Model Intercomparison." *J. Clim.*, 19, 3337–53.

Friedlingstein, P., M. Meinshausen, V. K. Arora, C. D. Jones, A. Anav, S. K. Liddicoat et al. 2014. "Uncertainties in CMIP5 Climate Projections Due to Carbon Cycle Feedbacks." *J. Clim.*, 27, 511–26.

Keenan, T. F., M. S. Carbone, M. Reichstein, and A. D. Richardson. 2011. "The Model-Data Fusion Pitfall: Assuming Certainty in an Uncertain World." *Oecologia*, 167, 587–97.

Keenan, T. F., E. A. Davidson, J. W. Munger, and A. D. Richardson. 2013. "Rate My Data: Quantifying the Value of Ecological Data for the Development of Models of the Terrestrial Carbon Cycle." *Ecol. Appl.*, 23, 273–86.

LeBauer, D., D. Wang, K. T. Richter, C. C. Davidson, and M. C. Dietze. 2013. "Facilitating Feedbacks between Field Measurements and Ecosystem Models." *Ecol. Monogr.*, 83, 133–54.

Luo, Y., T. F. Keenan, and M. Smith. 2014. "Predictability of the Terrestrial Carbon Cycle." *Glob. Chang. Biol.*, 21 (5), 1737–51.

Luo, Y., E. Weng, X, Wu, C. Gao, X. Zhou, and L. Zhang. 2009. "Parameter Identifiability, Constraint, and Equifinality in Data Assimilation with Ecosystem Models." *Ecol. Appl.*, 19, 571–74.

Matheny, A. M., G. Bohrer, P. C. Stoy, I. Baker, A. Black, A. R. Desai et al. 2014. "Characterizing the Diurnal Patterns of Errors in the Prediction of Evapotranspiration by Several Land-Surface Models: An NACP Analysis." *J. Geophys. Res. Biogeosci.*, 119 (7), 1458–73.

Matthes, J. H., C. Sturtevant, J. Verfaillie, S. Knox, and D. Baldocchi. 2014. "Parsing the Variability in $CH_4$ Flux at a Spatially Heterogeneous Wetland: Integrating Multiple Eddy Covariance Towers with High-resolution Flux Footprint Analysis." *J. Geophys. Res. Biogeosci.*, 119 (7), 1322–39.

Medlyn, B. E., R. A. Duursma, D. Eamus, D. S. Ellsworth, I. C. Prentice, C.V.M. Barton et al. 2011. "Reconciling the Optimal and Empirical Approaches to Modelling Stomatal Conductance." *Glob. Chang. Biol.*, 17, 2134–44.

Medvigy, D. M., S. C. Wofsy, J. W. Munger, D. Y. Hollinger, and P. R. Moorcroft. 2009. "Mechanistic Scaling of Ecosystem Function and Dynamics in Space and Time: Ecosystem Demography Model Version 2." *J. Geophys. Res.*, 114, 1–21.

Naithani, K. J., R. Kennedy, K. J. Davis, K. Keller, D. Bladwin, J. Masek et al. 2015. "Understanding and Quantifying Uncertainties in Upscaling $CO_2$ fluxes and Its Implications for Carbon Management." Plenary talk, NACP PI Meeting, January 27, Washington, DC.

Niu, S., Y. Luo, M. C. Dietze, T. Keenan, and Z. Shi. 2014. "The Role of Data Assimilation in Predictive Ecology." *Ecosphere*, 5, 1–16.

Oleson, K.W., D. M. Lawrence, B. Gordon, B. Drewniak, M. Huang, C. D. Koven et al. 2013. *Technical Description of Version 4.5 of the Community Land Model (CLM)*. Technical Note NCAR/TN-503+STR. Boulder, CO: National Center for Atmospheric Research.

Safta, C., D. Ricciuto, K. Sargsyan, B. Debusschere, H. N. Najm, M. Williams et al. 2014. "Global Sensitivity Analysis, Probabilistic Calibration, and Predictive Assessment for the Data Assimilation Linked Ecosystem Carbon Model." *Geosci. Model Dev. Discuss.*, 7, 6893–6948.

Schaefer, K. M., D. Y, Hollinger, E. Humphreys, B. Poulter, B. Raczka, A. D. Richardson et al. 2012. "A Model-Data Comparison of Gross Primary Productivity: Results from the North American Carbon Program Site Synthesis." *J. Geophys. Res.*, 117, 1–15.

Shiklomanov, A.A.N., M. C. Dietze, T. Viskari, and P. A. Townsend. 2016. "Quantifying the Influences of Spectral Resolution on Uncertainty in Leaf Trait Estimates through a Bayesian Approach to RTM Inversion." *Remote Sens. Environ.*, 183, 1–42.

Wang, D., D. S. Lebauer, and M. C. Dietze. 2013. "Predicted Yields of Short-rotation Hybrid Poplar (*Populus* spp.) for the Contiguous US." *Ecol. Appl.*, 23 (4), 944–58.

Weng, E., and Y. Luo. 2011. "Relative Information Contributions of Model vs. Data to Short- and Long-Term Forecasts of Forest Carbon Dynamics. *Ecol. Appl.*, 21, 1490–1505.

Williams, M., A. D. Richardson, M. Reichstein, P. C. Stoy, P. Peylin, H. Verbeeck et al. 2009. "Improving Land Surface Models with FLUXNET Data." *Biogeosciences*, 6, 1341–59.

Xia, J. Y., Y. Luo, Y.-P. Wang, E. S. Weng, and O. Hararuk. 2012. "A Semi-analytical Solution to Accelerate Spin-up of a Coupled Carbon and Nitrogen Land Model to Steady State." *Geosci. Model Dev.*, 5, 1259–71.

Zobitz, J. M., A. R. Desai, D.J.P. Moore, and M. A. Chadwick. 2011. "A Primer for Data Assimilation with Ecological Models using Markov Chain Monte Carlo (MCMC)." *Oecologia*, 167, 599–611.

## 13. DATA ASSIMILATION 1: ANALYTICAL METHODS

Caswell, H. 2001. *Matrix Population Models*. 2nd ed. Sunderland, MA: Sinauer Associates.

Dietze, M. C., D. S. LeBauer, and R. Kooper. 2013. "On Improving the Communication between Models and Data." *Plant Cell Environ.*, 36, 1575–85.

Fox, A., M. Williams, A. D. Richardson, D. Cameron, J. H. Gove, T. Quaife et al. 2009. "The REFLEX Project: Comparing Different Algorithms and Implementations for the Inversion of a Terrestrial Ecosystem Model against Eddy Covariance Data." *Agricult. Forest Meteorol.*, 149, 1597–1615.

Press, W. H., S. A. Teukolsky, W. T. Vettering, and B. P. Flannery. 2007. *Numerical Recipes: The Art of Scientific Computing*. 3rd ed. Cambridge: Cambridge University Press.

## 14. DATA ASSIMILATION 2: MONTE CARLO METHODS

Del Moral, P., A. Doucet, and A. Jasra. 2006. "Sequential Monte Carlo Samplers." *J. R. Stat. Soc. B*, 68, 411–36.

Doucet, A., S. Godsill, and C. Andrieu. 2000. "On Sequential Monte Carlo Sampling Methods for Bayesian Filtering." *Stat. Comput.*, 10, 197–208.

Doucet, A., and A. Johansen. 2011. "A Tutorial on Particle Filtering and Smoothing: Fifteen Years Later." *Handb. Nonlinear Filter.*, 12, 656–704.

Evensen, G. 2009a. *Data Assimilation: The Ensemble Kalman Filter*. 2nd ed. New York: Springer.

———.2009b. "The Ensemble Kalman Filter for Combined State and Parameter Estimation." *IEEE Control Syst. Mag.*, 29 (3), 83–104.

Gove, J. H., and D. Y. Hollinger. 2006. "Application of a Dual Unscented Kalman Filter for Simultaneous State and Parameter Estimation in Problems of Surface-Atmosphere Exchange." *J. Geophys. Res.*, 111, 1–21.

Hastie, D. I., and P. J. Green. 2012. "Model Choice Using Reversible Jump Markov Chain Monte Carlo." *Stat. Neerl.*, 66, 309–38.

Hoeting, J. A., D. Madigan, A. E. Raftery, and C. T. Volinsky. 1999. "Bayesian Model Averaging: A Tutorial." *Stat. Sci.*, 14, 382–417.

Julier, S., J. Uhlmann, and H. F. Durrant-Whyte. 2000. "A New Method for the Nonlinear Transformation of Means and Covariances in Filters and Estimators." *IEEE Trans. Automat. Contr.*, 45, 477–82.

Lee, J.-B., M.-J. Kim, S, Yoon, and E.-Y. Chung. 2012. "Application-Support Particle Filter for Dynamic Voltage Scaling of Multimedia Applications." *IEEE Trans. Comput*, 61, 1256–69.

Raftery, A. E., T. Gneiting, F. Balabdaoui, and M. Polakowski. 2005. "Using Bayesian Model Averaging to Calibrate Forecast Ensembles." *Mon. Weather Rev.*, 133, 1155–74.

Schwalm, C. R., C. A. Williams, K. Schaefer, R. Anderson, M. A. Arain, I. Baker et al. 2010. "A Model-Data Intercomparison of $CO_2$ Exchange across North America: Results from the North American Carbon Program Site Synthesis." *J. Geophys. Res.*, 115 (G3), G00H05.

Tebaldi, C., and R. Knutti. 2007. "The Use of the Multi-model Ensemble in Probabilistic Climate Projections." *Philos. Trans. A. Math. Phys. Eng. Sci.*, 365, 2053–75.

Uhlmann, J. K. 1995. "Dynamic Map Building and Localization: New Theoretical Foundations." PhD dissertation, University of Oxford, Department of Engineering Science.

Weigel, A. P., M. A. Liniger, and C. Appenzeller. 2007. "Can Multi-model Combination Really Enhance the Prediction Skill of Probabilistic Ensemble Forecasts?" *Quar. J. R. Meteorol. Soc.*, 133, 937–48.

## 15. EPIDEMIOLOGY

Anderson, R. M., and R. M. May. 1983. "Vaccination against Rubella and Measles: Quantitative Investigations of Different Policies." *J. Hyg. (Lond.)*, 90, 259–325.

Bharti, N., A. J. Tatem, M. J. Ferrari, R. F. Grais, A. Djibo, and B. T. Grenfell. 2011. "Explaining Seasonal Fluctuations of Measles in Niger Using Nighttime Lights Imagery." *Science*, 334, 1424–27.

Bjørnstad, O. N., B. F. Finkenstadt, and B. T. Grenfell. 2002. "Dynamics of Measles Epidemics: Estimating Scaling of Transmission Rates Using a Time Series SIR Model." *Ecol. Monogr.*, 72, 169–84.

Chen, S., J. Fricks, and M. J. Ferrari. 2012. "Tracking Measles Infection through Non-linear State Space Models." *J. R. Stat. Soc. C*, 61, 117–34.

Diamond, J. M. 1999. *Guns, Germs, and Steel: The Fates of Human Societies*. New York: W. W. Norton.

Dietze, M. C., and J. H. Matthes. 2014. "A General Ecophysiological Framework for Modeling the Impact of Pests and Pathogens on Forest Ecosystems." *Ecol. Lett.*, 17 (11), 1418–26.

Ellner, S. P., B. A. Bailey, G. V. Bobashev, A. R. Gallant, B. T. Grenfell, and D. W. Nychka. 1998. "Noise and Nonlinearity in Measles Epidemics: Combining Mechanistic and Statistical Approaches to Population Modeling." *Am. Nat.*, 151, 425–40.

Ferguson, N. M., C. A. Donnelly, and R. M. Anderson. 2001a. "The Foot-and-Mouth Epidemic in Great Britain: Pattern of Spread and Impact of Interventions." *Science*, 292, 1155–60.

Ferguson, N. M., C. A. Donnelly, and R. M. Anderson. 2001b. "Transmission Intensity and Impact of Control Policies on the Foot and Mouth Epidemic in Great Britain." *Nature*, 413, 542–48.

Ferrari, M. J., R. F. Grais, N. Bharti, A.J.K. Conlan, O. N. Bjørnstad, L. J. Wolfson et al. 2008. "The Dynamics of Measles in Sub-Saharan Africa." *Nature*, 451, 679–84.

Grenfell, B., O. N. Bjørnstad, and B. F. Finkenstadt. 2002. "Dynamics of Measles Epidemics: Scaling Noise, Determinism, and Predictability with the TSIR Model." *Ecol. Monogr.*, 72, 185–202.

Grenfell, B. T., O. N. Bjørnstad, and J. Kappey. 2001. "Travelling Waves and Spatial Hierarchies in Measles Epidemics." *Nature*, 414, 716–23.

Grube, A., D. Donaldson, T. Keily and L. Wu. 2011. *Pesticides Industry Sales and Usage: 2006 and 2007 Market Estimates*. Washington, DC: U.S. Environmental Protection Agency.

Hatala, J., M. C. Dietze, R. Crabtree, D. Six, K. Kendall, and P. Moorcroft. 2011. "An Ecosystem-Scale Model for the Spread of a Host-Specific Forest Pathogen in the Greater Yellowstone Ecosystem." *Ecol. Appl.*, 21, 1138–53.

Hicke, J. A., C. D. Allen, A. R. Desai, M. C. Dietze, R. J. Hall, E. H. Hogg et al. 2012. "Effects of Biotic Disturbances on Forest Carbon Cycling in the United States and Canada." *Glob. Chang. Biol.*, 18, 7–34.

Keeling, M. J., P. Rohani, and B. T. Grenfell. 2001a. "Seasonally Forced Disease Dynamics Explored as Switching between Attractors." *Physica D*, 148, 317–35.

Keeling, M. J., M. E. Woolhouse, D. J. Shaw, L. Matthews, M. Chase-Topping, D. T. Haydon et al. 2001b. "Dynamics of the 2001 UK Foot and Mouth Epidemic: Stochastic Dispersal in a Heterogeneous Landscape." *Science*, 294, 813–17.

Keeling, M. J., M.E.J. Woolhouse, R. M. May, G. Davies, and B. T. Grenfell. 2003. "Modelling Vaccination Strategies against Foot-and-Mouth Disease." *Nature*, 421, 136–42.

Kermack, W. O., and A. G. McKendrick. (1927). "A Contribution to the Mathematical Theory of Epidemics." *Proc. R. Soc. A Math. Phys. Eng. Sci.*, 115, 700–721.

Kitching, R. P., A. M. Hutber, and M. V. Thrusfield. 2005. "A Review of Foot-and-Mouth Disease with Special Consideration for the Clinical and Epidemiological Factors Relevant to Predictive Modelling of the Disease." *Vet. J.*, 169, 197–209.

LaDeau, S. L., G. E. Glass, N. T. Hobbs, A. Latimer, and R. S. Ostfeld. 2011. "Data-Model Fusion to Better Understand Emerging Pathogens and Improve Infectious Disease Forecasting." *Ecol. Appl.*, 21, 1443–60.

Meentemeyer, R. K., S. E. Haas, and T. Václavík. 2012. "Landscape Epidemiology of Emerging Infectious Diseases in Natural and Human-Altered Ecosystems." *Annu. Rev. Phytopathol.*, 50, 379–402.

Ong, J.B.S., M.I.-C. Chen, A. R. Cook, H. C. Lee, V. J. Lee, R.T.P. Lin et al. 2010. "Real-time Epidemic Monitoring and Forecasting of H1N1-2009 Using Influenza-like Illness from General Practice and Family Doctor Clinics in Singapore." *PLoS One*, 5, e10036.

Pimentel, D. 2005. "Environmental and Economic Costs of the Application of Pesticides Primarily in the United States." *Environ. Dev. Sustain.*, 7, 229–52.

Thompson, D., P. Muriel, D. Russell, P. Osborne, A. Bromley, M. Rowland et al. 2002. "Economic Costs of the Foot and Mouth Disease Outbreak in the United Kingdom in 2001." *Rev. Sci. Tech.*, 21, 675–87.

Tildesley, M. J., R. Deardon, N. J. Savill, P. R. Bessell, S. P. Brooks, M.E.J. Woolhouse et al. 2008. "Accuracy of Models for the 2001 Foot-and-Mouth Epidemic." *Proc. Biol. Sci.*, 275, 1459–68.

Tildesley, M. J., N. J. Savill, D. J. Shaw, R. Deardon, S. P. Brooks, M.E.J. Woolhouse et al. 2006. "Optimal Reactive Vaccination Strategies for a Foot-and-Mouth Outbreak in the UK." *Nature*, 440, 83–86.

WHO. (2004). *The World Health Report 2004: Changing History*. Geneva: World Health Organization.

———. (2013). *World Health Statistics*. Geneva: World Health Organization.

## 16. ASSESSING MODEL PERFORMANCE

Blyth, E., D. B. Clark, R. Ellis, C. Huntingford, S. Los, M. Pryor, M. Best, and S. Sitch. 2011. "A Comprehensive Set of Benchmark Tests for a Land Surface Model of Simultaneous Fluxes of Water and Carbon at Both the Global and Seasonal Scale." *Geosci. Model Devel.*, 4, 255–69.

Bröcker, J., and L. A. Smith. 2007. "Scoring Probabilistic Forecasts: The Importance of Being Proper." *Weather Forecast.*, 22, 382–88.

Dietze, M. C., R. Vargas, A. D. Richardson, P. C. Stoy, A. G. Barr, R. S. Anderson et al. 2011. "Characterizing the Performance of Ecosystem Models across Time Scales: A Spectral Analysis of the North American Carbon Program Site-level Synthesis." *J. Geophys. Res.*, 116 (G4), G04029.

Hoffman, F. M., J. W. Larson, R. T. Mills, B.-G.J. Brooks, A. R. Ganguly, W. W. Hargrove et al. 2011. "Data Mining in Earth System Science (DMESS 2011)." *Proc. Computer Sci.*, 4, 1450–55.

Hyndman, R. J., and A. B. Koehler. 2006. "Another Look at Measures of Forecast Accuracy." *Int. J. Forecast.*, 22, 679–88.

Luo, Y., J. T. Randerson, G. Abramowitz, C. Bacour, E. Blyth, N. Carvalhais et al. 2012. "A Framework for Benchmarking Land Models. *Biogeosciences*, 9, 3857–74.

Medlyn, B. E., S. Zaehle, M. G. De Kauwe, A. P. Walker, M. C. Dietze, P. J. Hanson et al. 2015. "Using Ecosystem Experiments to Improve Vegetation Models." *Nature Climate Change*, 5, 528–34.

Milly, P.C.D., J. Betancourt, M. Falkenmark, R. M. Hirsch, Z. W. Kundzewicz, D. P. Letten-maier, and R. J. Stouffer. 2008. "Stationarity Is Dead: Whither Water Management?" *Science*, 319, 573–74.

Schaefer, K. M., D. Y. Hollinger, E. Humphreys, B. Poulter, B. Raczka, A. D. Richardson et al. 2012. "A Model-Data Comparison of Gross Primary Productivity: Results from the North American Carbon Program Site Synthesis." *J. Geophys. Res.*, 117, 1–15.

Stoy, P. C., M. C. Dietze, A. D. Richardson, R. Vargas, A. G. Barr, R. S. Anderson, et al. 2013. "Evaluating the Agreement between Measurements and Models of Net Ecosystem Exchange at Different Times and Time Scales Using Wavelet Coherence: An Example Using Data from the North American Carbon Program Site-Level Interim Synthesis." *Biogeosciences*, 10, 6893–6909.

Walker, A. P., P. J. Hanson, M. G. De Kauwe, B. E. Medlyn, S. Zaehle, S. Asao et al. 2014. "Comprehensive Ecosystem Model Data Synthesis Using Multiple Data Sets at Two Temperate Forest Free Air $CO_2$ Enrichment Experiments: Model Performance at Ambient $CO_2$ Concentration." *J. Geophys. Ress.—Biogeosci.*, 119, 937–64.

Wenger, S. J., and J. D. Olden. 2012. "Assessing Transferability of Ecological Models: An Underappreciated Aspect of Statistical Validation." *Meth. Ecol. Evol.*, 3 (2), 260–67.

## 17. PROJECTION AND DECISION SUPPORT

Berger, J. O. 1985. *Statistical Decision Theory and Bayesian Analysis*. 2nd ed. New York: Springer.

Biggs, R., M. W. Diebel, D. Gilroy, A. M. Kamarainen, M. S. Kornis, N. D. Preston et al. 2010. "Preparing for the Future: Teaching Scenario Planning at the Graduate Level." *Front. Ecol. Environ.*, 8, 267–73.

Clark, J. S., S. R. Carpenter, M. Barber, S. Collins, A. Dobson, J. A. Foley et al. 2001. "Ecological Forecasts: An Emerging Imperative." *Science*, 293, 657–60.

Fingerhut, B. P., W. Zinth, and R. de Vivie-Riedle. 2010. "The Detailed Balance Limit of Photochemical Energy Conversion." *Phys. Chem. Chem. Phys.*, 12, 422–32.

Gregory, R., L. Failing, M. Harstone, G. Long, T. McDaniels, and D. Ohlson. 2012. *Structured Decision Making: A Practical Guide to Environmental Management Choices*. New York: Wiley.

Kahneman, D. 2013. *Thinking Fast and Slow*. New York: Farrar, Straus and Giroux.

MacCracken, M. 2001. "Prediction versus Projection—Forecast versus Possibility." *Weather-Zine*, 26, 3–4.

Moss, R. H., J. A. Edmonds, K. A. Hibbard, M. R. Manning, S. K. Rose, D. P. van Vuuren et al. 2010. "The Next Generation of Scenarios for Climate Change Research and Assessment. *Nature*, 463, 747–56.

Peterson, G. D., G. S. Cumming, and S. R. Carpenter. 2003. "Scenario Planning: A Tool for Conservation in an Uncertain World." *Conserv. Biol.*, 17, 358–66.

Regan, H. M., M. Colyvan, and M. A. Burgman. 2002. "A Taxonomy and Treatment of Uncertainty for Ecology and Conservation Biology." *Ecol. Appl.*, 12, 618–28.

Schwartz, B. 2005. *The Paradox of Choice: Why Less Is More*. New York: Harper Perennia.

Taleb, N. N. 2007. *The Black Swan: The Impact of the Highly Improbable*. 2nd ed. New York: Random House.

United Nations. 1993. *Agenda 21: Programme of Action for Sustainable Development; Rio Declaration on Environment and Development; Statement of Forest Principles: The Final Text of Agreements Negotiated by Governments at the United Nations Conference on Environment and Development (UNCED), 3–14 June 1992, Rio De Janeiro, Brazil*. New York: United Nations, Department of Public Information.

Walters, C. 1986. *Adaptive Management of Renewable Resources*. New York: Macmillan.

## 18. FINAL THOUGHTS

Clark, J. S. 2003. "Uncertainty and Variability in Demography and Population Growth: A Hierarchical Approach." *Ecology*, 84, 1370–81.

Clark, J. S., and O. N. Bjørnstad. 2004. "Population Time Series: Process Variability, Observation Errors, Missing Values, Lags, and Hidden States." *Ecology*, 85, 3140–50.

Clark, J. S., A. E. Gelfand, C. W. Woodall, and K. Zhu. 2014. "More Than the Sum of the Parts: Forest Climate Response from Joint Species Distribution Models." *Ecol. App.*, 24, 990–99.

LeBauer, D., D. Wang, K. T. Richter, C. C. Davidson, and M. C. Dietze. 2013. "Facilitating Feedbacks between Field Measurements and Ecosystem Models." *Ecol. Monogr.*, 83, 133–54.

Moorcroft, P. R., G. C. Hurtt, and S. W. Pacala. 2001. "A Method for Scaling Vegetation Dynamics: The Ecosystem Demography Model (ED)." *Ecol. Monogr.*, 71, 557–86.

Petchey, O. L., M. Pontarp, T. M. Massie, S. Kéfi, A. Ozgul, M. Weilenmann et al. 2015. "The Ecological Forecast Horizon, and Examples of Its uses and Determinants." *Ecol. Lett.*, 18 (7), 597–611.

Turchin, P. 2003. *Complex Population Dynamics: A Theoretical/Empirical Synthesis (MPB-35)*. Princeton, NJ: Princeton University Press.

Turner, M., W. Romme, and R. Gardner. 1993. "A Revised Concept of Landscape Equilibrium: Disturbance and Stability on Scaled Landscapes. *Landscape Ecol.*, 8, 213–27.

# Index